AF531293

PLANT BIOTECHNOLOGY

PLANT BIOTECHNOLOGY

By

Ashok Kumar

Lecturer
Department of Zoology
Bundelkhand University
Campus Department
Jhansi

DISCOVERY PUBLISHING HOUSE
NEW DELHI-110002

Reprinted - 2019

First Published - 2006

ISBN: 978-81-8356-132-7

Plant Biotechnology

Published by:
DISCOVERY PUBLISHING HOUSE PVT. LTD.
4383/4B, Ansari Road, Darya Ganj
New Delhi-110 002 (India)
Phone: +91-11-23279245, 23253475; 43596065
E-mail: discoverybooksindia@gmail.com
discoverypublishinghouse@gmail.com
web: www.discoverypublishinggroup.com

Printed at:
Infinity Imaging Systems
Delhi

Preface

The present title **Plant Biotechnology** offers an engaging introduction to this exciting area of scientific research and technology. It has been written for advanced undergraduate, post-graduate students and other researchers in this area taking a broad range of cources. The present title also provides an overview of the production of GM crops, highlighting the key scientific and technical advances that underpin their development. It takes the students from the basic of the subject through to the latest thinking.

The aim of this book is to provide enough information and examples to give the reader a sound knowledge of plant biotechnology in all its various guises, but particularly those related to the genetic manipulation to crop plants. It is not intended to provide an encyclopaedic coverage of the subject; such a task would require a volume far larger than this. As such, this is a textbook, and therefore a learning and teaching aid, rather than an academic treatise. Hopefully, this book also encourages a critical appraisal of plant biotechnology—not just the scientific aspects, but also the economic, social, moral and ethical issues that surround, and some would say, plague the subject.

There can be no claim to originality except in the manner of treatment and much of the information has been obtained from the books and scientific journals available in the different libraries.

The author expresses his thanks to his friends and colleagues whose continue inspirations have initiated him to bring out this book.

The author is painfully aware of the shortcomings, errors and misprints that have crept in, and shall be greateful to receive suggestion for improvement of the next edition from all the readers.

The author expresses his gratitude to Mr. Wasan and staff of M/s Discovery Publishing House for their whole hearted co-operation in the publication of this book.

Author

CONTENTS

Chapter **1**

INTRODUCTION **1—17**

Rye, Millet, Sorghum, Plant Breeding, History of Plant Breeding, Modern Plant Breeding, Breeding Strategies for In-Breeding Crops, Breeding Strategies in Out-Breeding Crops, Genetic Diversity, Concluding Remarks

Chapter **2**

PRODUCT DEVELOPMENT **18—42**

Commercial Targets, Problems in Cereal Biotechnology, Commercial Traits, Molecular Breeding of Transgenic Plants, Molecular Quality Control for Transgenic Plants, Intellectual Property, Regulatory Issues, Marketing Strateiges, Future Trends.

Chapter **3**

BIOTECHNOLOGY VALUE TO CEREALS **43—58**

Weed Control, Disease Resistance, Improved Nutritional Properties, Improved Processing Properties, Improved Cereal Quality Control.

Chapter **4**

MOLECULAR BIOLOGICAL TOOLS IN PLANT BREEDING **59—85**

Plant Breeding, Markers, Characters, Deployment of Molecular Markers, Future Prospects, Conclusions.

Chapter 5

ASSESMENT AND LEGISLATIVE ISSUES **86—116**

GM Crop Development, Risk Management and Avoidance, Impact of Genetically Modified Crops, How is Biotechnology Regulated?, Public Perceptions, Future Developments in the Regulatory Process.

Chapter 6

GENETICALLY IMPROVED FOOD **117—132**

Changing the Nutritional Value of Foods, Modification of Fruit Colour and Sweetness, Modification of Food-processing, Molecular Farming, Future Trends.

Chapter 7

NUTRITIONAL ENHANCEMENT OF PLANT FOODS **133—147**

The Nutritional Importance of Plants, Strategies for Nutritional Enhancement, Nutritional Enhancement, Bioavailability, Nutritional Enhancement, Constraints on Innovation.

Chapter 8

STRESS TOLERANCE **148—166**

The Nature of Abiotic Stress, The Nature of Water-deficit Stress, Manipulation of Tolerance to Specific Water-defic it Stresses, Secondary Effects of Abiotic Stress.

Chapter 9

MOLECULAR FARMING **167—204**

Carbohydrates and Lipids, Cyclodextrins from Starch, Trehalose, Improvement of Plant Oils, Production of Shorter Chain Fatty Acids, Production of Longer Chain Fatty Acids, Modification of the Degree of Saturation, Production of Rare Fatty Acids, Molecular Farming of Protein, Strategies for Protein Production, Stable Expression, Transient Expression, The production of Enzymes for Industrial Uses, Antibodies or 'Plantibodies', Plant-derived Vaccines, Other Proteins—Biopharmaceuticals, Trichosanthin, Glucocerebrosidase, Human Serum Albumin, Arginine Vasopressin, Economic Considerations.

Chapter 10

GENETIC IMPROVEMENT OF CROPS **205—237**

Genes are Made of DNA, Multipartite Nature of Genes, Manipulation of Genes in the Laboratory, Plant Regeneration Technology, Introduction of DNA Using a "Gene Gun", Genomics Improvement of Crop.

Chapter 11

GENES IN GROWTH AND DEVELOPMENT **238—272**

Cells, Tissues, and Organs, Plant Development by Repetitive Organ Formation, Environmental Stimuli and Hormones Plant Development, Development of Seeds, Formation of the Vegetative Body, Reproductive Development of Flowers, The Formation of Fruits Aids Seed Dispersal, Integral Processes of Development, Hormonal Signals to the Nucleus to Regulate Gene Activity, Genetic Modification of Crops.

Chapter 12

SEEDS TECHNOLOGY IN AGRICULTURE **273—308**

Nutritional Reserves to Support the Growing Seedling, Seed Maturation, Seed Germination, Seed Production is Often Distinct from Crop Production, Seed Certification Programs, Seed Banks, Seeds are the Major Delivery System for Plant Biotechnology, Terminator Gene Technology, Seed's Ability to Function as a Propagule.

INDEX **309**

1

Introduction

Cereals owe their English name to the Roman goddess Ceres, the giver of grain, indicative of the antiquity and importance of cereals. This importance is still very much the case today; cereals of one sort or another sustain the bulk of mankind's basic nutritional needs, both directly and indirectly as animal feed. It is primarily the grains of cereals that are useful to us, although the vegetative parts of the plant may be used as fodder or for silage production, and straw is used for animal bedding.

Cereals are members of the large monocotyledonous grass family, the *Gramineae*. The flowering organs are carried on a stem called the rachis, which may be branched, and in turn bears spikelets which may carry more than one flower at each node of the rachilla. The spikelets may be organized in a loose panicle as in sorghum, oats and some millets, or in a tight spike, as in wheat. The length of the internodes of the rachis and of the rachilla, and the number of flowers at each node of the spikelet determine the overall architecture. Each spikelet is subtended by two bracts or leaf-like organs termed the glumes, and each flower is the spikelet is enclosed in two bract-like organs called the lemma and palea. The lemma may be extended to form a long awn. In some cereals or cereal varieties the lemma and palea may remain attached to the grain; these are termed hulled or husked grains, such as oats and most barleys, as opposed to naked grains such as most wheas ans maize.

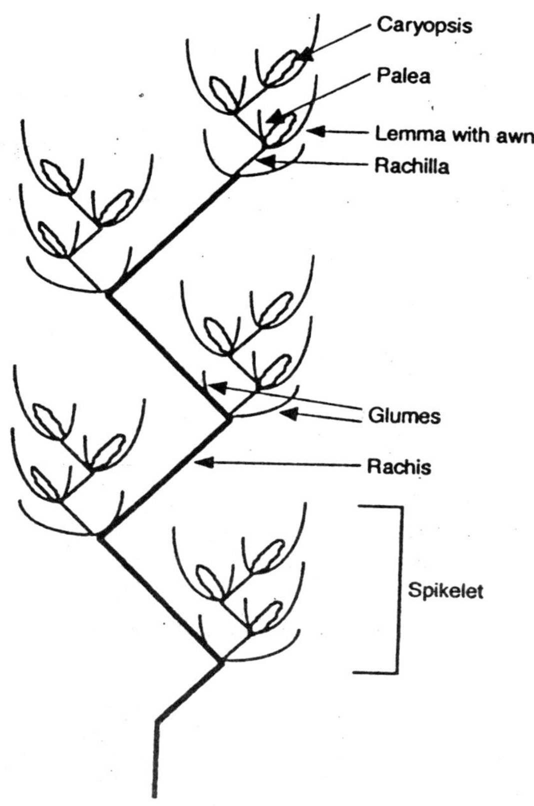

Figure 1.1: ***Generalised structure of cereal flowering organs. The length and branching pattern of the rachis and the rachilla, and the number of flowers per spikelets determine the overall appearance of the cereal.***

The cereals, with the exception of maize, are dioecious. Each flower bears both male organs; the three anthers (six in rice), and female organs; the ovary which carries two feathery stigmas. In maize, the male flowers are borne in spikes on a terminal panicle called a tassel, and the female flowers are in spikelets borne in rows on the swollen tips of lateral branches, the cobs. The main storage organ of cereal seeds in the endosperm which makes up the bulk of the grain, and primarily consists of starch and protein. The grain is botanically a fruit known as a caryopsis; in this structure, the wall of the seed (the testa) becomes fused with the maternally derived ovary wall (the pericarp).

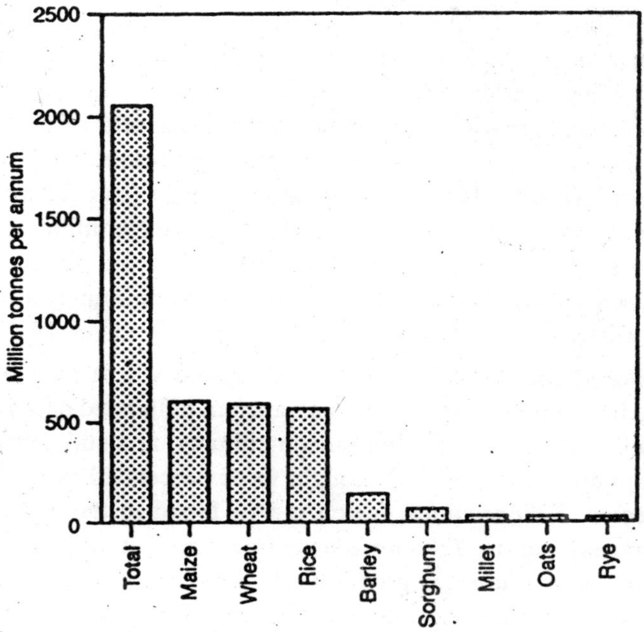

Figure 1.2: ***Annula global production of cereals in millions of tonnes.***

Cereals have developed their importance as food plants because they are high yielding, with world average yields around three tonnes per hectare. The grains are very nutritious; generalized cereal grain contents (which will of course vary with species, growing conditions and variety) are: carbohydrates (70%), protein (10%), lipids (3%). Being dessicated at harvest with a water content of about 12%, cereal grains are easy and economical to transport and store. Different cereals have risen to eminence in different quarters of the globe because of geographical provenance and because of differing climatic and environmental requirements for growth, but their shared favourable characteristics underline their importance as staple foodstuffs. Three cereals - wheat, maize and rice - make up the bulk of world cereals production, but five other cereal crops also make important contributions to world nutrition, and to food and drink production. In order of global production tonnage, these are barley, sorghum, millets, oats and rye.

Wheat

Wheat is an ancient cultivated crop, whose origins are not clear, but most of the evidence points towards the Middle East as the geographical region of origin. There are three sets of wheat species, differing in ploidy (basic chromosome number). *Triticum monococcum* (inkorn) is a 'primitive' diploid species (haploid chromosome number 7), whose use goes back to the Neolithic, and which is still cultivated to some extent in Europe. *Triticum boeoticum* is a wild form of *T.monococcum*, to be found in the Balkans and eastern Mediterranean.

Triticum dicoccum (emmer) is a tetraploid wheat (haploid chromosome number 14), and also an ancient cultivated species, associated with the old Mediterranean cultures, and still grown in some parts of Europe. It is thought to be descended from the wild species, *T.dicoccoides*, which is still found in the eastern Mediterranean region. *Triticum durum* (macaroni wheat), in turn descended from emmer, is grown world wide and has excellent pasta-making qualities. *Triticum timopheevi*, *T. turgidum* (poulard, rivet or cone wheat), *T.turanicum* (khorasan wheat), *T. polonicum* (Polish wheat or giant rye) and *T.carthlicum* (Persian wheat) a[illegible] some species of cultivated tetraploid wheat, but now of relativel[illegible] minor economic importance.

Triticum aestivum is the hexaploid wheat (haploid chromosome number 21) and of all the wheats, this is most commonly grown today. It is thought that diploid einkorn and tetraploid emmer wheats may be ancestral to modern hexaploid wheats. No wild hexaploid species are known, but there are several cultivated subspecies, previously considered by some authorities to be separate species (subsp. *spelta* (spelt or dinkel), *macha, vavilovii, vulgare* (bread wheat), *compactum* (club wheat), and *sphaero-*

coccum (shot wheat). The most widespread hexaploid wheat grown today is the bread wheat *Triticum aestivum* subsp. *vulgare.* Winter wheats are sown in autumn, vernalise over winter (vernalisation is a cold treatment required to induce flowering) and are harvested in early summer. Spring wheats are sown in spring and harvested in late summer, they generally have a lower yield than winter wheats.

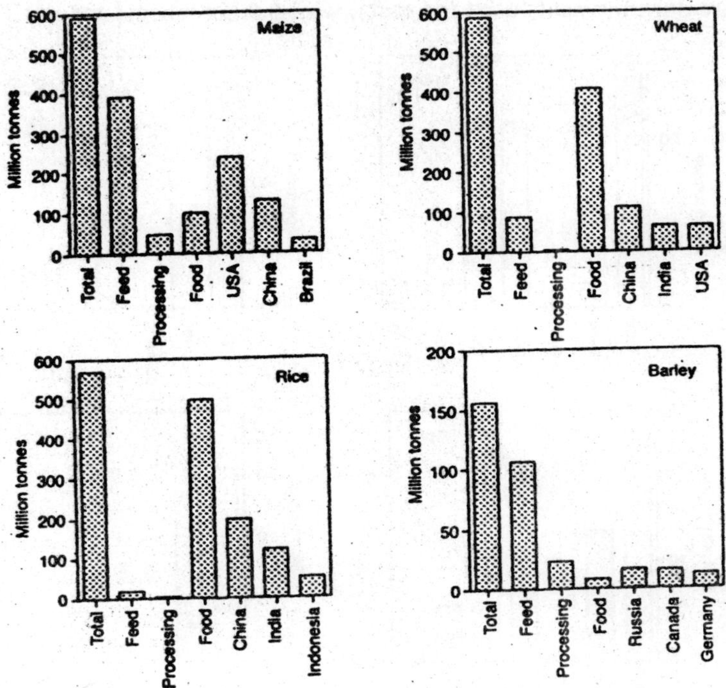

Figure 1.3 : ***Annual global production and utilisation of the eight most important cereal crops. Total production, utilization for animal feed, processing (industrial uses and processed foods), and direct huma[illegible] sumption, and the three largest producers are shown.***

Wheat is one of the most widely grown cereals, accounting for over one-quarter of the world's global cereal production, and is primarily used for human consumption with some 15% being used for animal feed. The largest global producers of wheat are China,

India and the USA. Wheats can be classified according to Kernel hardness: the distinction between hard and soft wheats was made even in Roman times. In American terminology, hard wheats with high protein to starch ratios (16% protein, 61% starch) make 'strong' flour, used in bread-making, whereas 'soft' wheats with 12% protein and 66% starch make 'weak' flours, used in biscuit manufacture. In continental European terminology, 'hard' wheats are durum wheats used for pasta, whilst other wheats are soft.

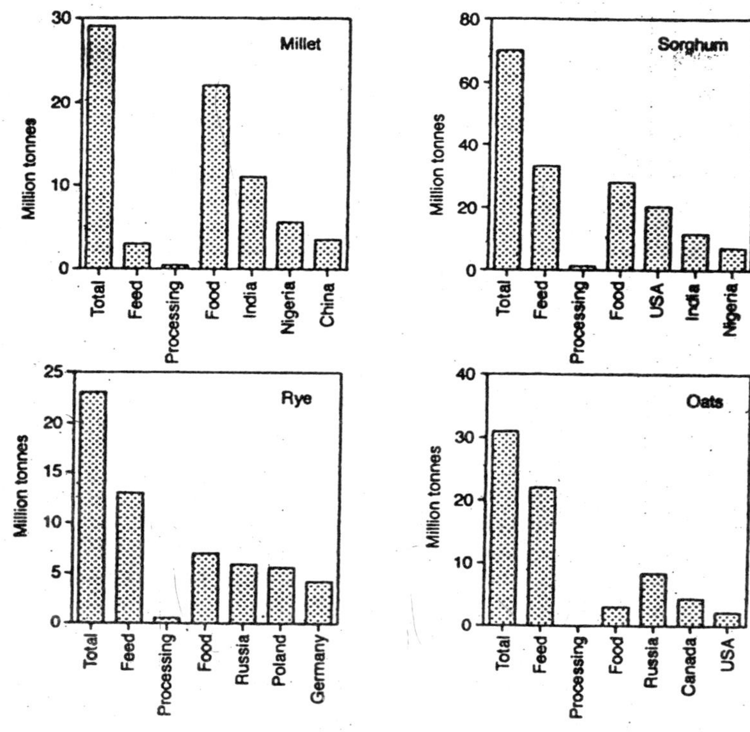

Figure 1.3 Continued

Maize

Maize (or corn in North America) (*Zea mays*) derives from and was domesticated in central America some 4000 years ago; a maize goddess, Cinteutl, was worshipped in Mexico. The true ancestor of maize is not known, but is shares a common ancestor with the weedy species teosinte (*Zea mexicana*).Maize is now grown throughout the world, the main producers being the USA, China and Brazil.

There are many maize subspecies with different agricultural uses, for example varieties *saccharata* (sweetcorn), *everta* (popcorn) *americana* (dent maize, grown in North America), *praecox* (flint maize, grown in Europe), *amylacea* (flour or soft maize, grown by American Indians) and *tunica* (podcorn). Maize accounts for over one-quarter of global cereal production, with the majority of the crop going for animal feed; however a substantial tonnage is used directly for human foods and for processing into manufactured foods, drinks and industrial raw materials.

Rice

The most commonly farmed species, *Oryza sativa*, is thought to have been domesticated in southern Asia some 6000 years ago, written evidence for the cultivation of rice (sometimes termed paddy) in China goes back to at least 2800 BC. Alexander the Great is said to have brought rice to Europe. The progenitor of domesticated rice is the wild species *Oryza rufipogon*. A second rice species (*Oryza glaberrima*) was domesticated in West Africa, and is still and important cultivated species in tropical Africa. Today, rice is a staple foodstuff of Asia and is grown throughout tropical and warm temperate regions. It is grown either immersed in water until harvest (the higher-yielding lowland rice) or on dry land (upland and hill rice). There are two main subspecies of *Oryza sativa*, the generally short-grained *japonica*, typically grown in more northern or southern regions with longer photoperiods, and the longer-grained *indica*, grown in more tropical regions. There are hard-and soft-grained (glutinous) varieties of both subspecies and many thousands of cultivated varieties. Rice accounts for over one-quarter of global cereal production, with the most majority going for human food. China, India and indonesia account for 65% of the world's production.

Barley

Before (*Hordeum vulgare*) is of an ancient lineage, being used for bread even before wheat in Neolithic times. It is thought to have arisen in south-western Asia or northern Africa and wild forms of two-row barley are still to be found in western Asia (*Hordeum spontaneum*). Barley is a crop of temperate climates and is a morphologically rather variable species, which has given taxonomists much employment, but in this chapter the view will be taken that there is one cultivated species with several subspecies (for example *distichon, hexastichon, agriocrithon, deficiens*). Two- and six-row barleys are the most commonly cultivated forms, six-row barley being more resistant to temperature extremes. The spike (or ear) consists of alternating nodes each bearing three spikelets (each a single flower). In two-row barley, only the central spikelets is fertile, but in six-row barley, all three spikelets are fertile. The ear may be erect or drooping at maturity, awned or awn-less. Winter barleys are sown in autumn, vernalized over winter and are harvested in early summer. Spring barleys are sown in spring and harvested in late summer. Barley accounts for some 7.5% of global cereal production. The majority of the barley crop is used as animal feedstuff, but about 15% is used for the production of beer and spirits. Russia, Canada and Germany were the world's biggest barley producers for 1996.

Oats

Oats have a long and uncertain pedigree, being known since early historical times, for example to the ancient Greeks. The crop was famously defined by Samuel Johnson in his dictionary as 'a grain, which in England is generally, given to horses, but is Scotland supports the people'. However as well as animal food (67% of the world crop), oats are widely used as human nutrition (10%) even outside of Scotland. The most important cultivated species is *Avena sativa*, but other species are also cultivated to a lesser extent, such as *Avena orientalis*, *A. nuda* and *A. brevis*. There are several wild species of oats, some of which may be ancestral to the cultivated oats, but today are troublesome weeds (for example, *Avena fatua*). Oats form a minor component of the world cereal crop at 1.5% of the total global cereal production and the biggest current producers are Russia, Canada and the USA.

RYE

Rye (*Secale cereale*) appears to be more recently domesticated than other cereal, although it was known to the ancient Greek and Roman civilization. The probable ancestor is *Secale montanum*, a wild species to be found in the Black and Caspian Sea areas. Rye is predominantly produced in central and eastern Europe. Both rye and oats may have originated as weed species in wheat and barley crops. Rye is a very winter-hardy crop that will grow on poor soils such as those of the north European plain. Rye accounts for only around 1% of world total cereal production , is used for animal food and human consumption and the prime producers are Russia, Poland and Germany (Figure 1.3).

MILLET

Millet is the collective name for a mumber of cereals species of importance as food crops in tropical and subtropical countries or as forage crops in more northern climates. These species include: *Eleusine coracana* (finger millet), *Setaria italica* (foxtail millet), *Echinochloa crus-galli* (Japanese barnyard millet), *Pennisetum glaucum* (pearl millet),and *Panicum miliaceum* (proso millet). Although generally low yielding, these crops are often grown in conditions under which other crops would not flourish. Millet forms only 1.5% of the total global cereal crop , and is primarily grown for food. The biggest producers are India, Nigeria and China.

SORGHUM

Sorghum (*Sorghum vulgare*) is a native of Africa and Asia and the many varieties which are cultivated there are important as human foods and as animal fodder. There are four general classes, grain sorghum, grass sorghum, broom corn and sweet sorghum or sorgo. Sorghum accounts for 3.5% of global cereal production , the primary producer being USA, India and Nigeria. In the USA, sorghum is mainly grown for animal fodder, but in India and Nigeria, the majority to the crop is used for human consumption.

PLANT BREEDING

World cereal production and yield per hectare have increased steadily over the last forty years. This trend is mirrored by the increased use of fertilisers and pesticides. However, much of the increase in yield and production can be attributed to improvements in crop varieties brought about by the efforts of plant breeders.

HISTORY OF PLANT BREEDING

During the course of plant domestication, the elements of plant breeding arose. Early agriculturalists would have taken an empirical approach to selecting their crops. As a result of this, domesticated crops differ from their wild progenitors in a number of important respects. Wild species disperse their seeds in order to spread their offspring far and wide. In wild cereals, the spikes bearing the

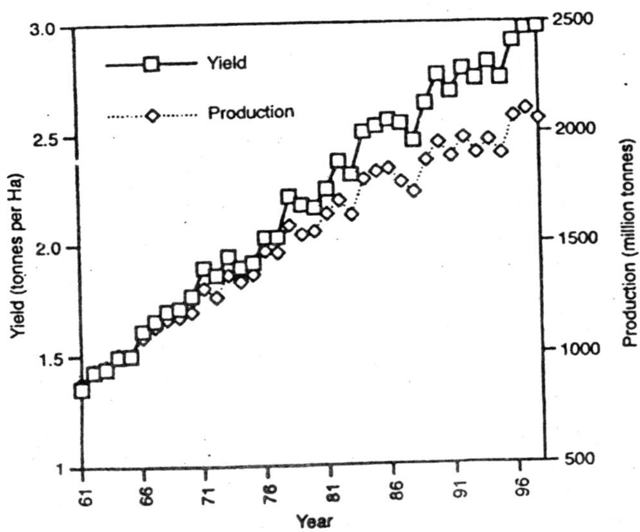

Figure 1.4: ***Annual global cereal production in millions of tonnes and cereal yield in tonnes per hectare for the year 1961-98.***

grains are borne on a brittle rachis (the main axis of the ear on which the grains are carried, and lose the grain if mechanically disturbed. In a cultivated crop, these seed would be lost prior to harvest, so non-dispersing crops with a rachis that did not easily shatter were 'selected' by man; these crops are thereshed after harvest. Domestication would also have put selection pressure on nondormant crop lines, since only those plants that germinated soon sowing would be included in the harvest. And of course, crops with enhanced yield and taste would be selected for by the early farmers.

This essentially informal process of selection by early farmers continued right through until recent times, and through time resulted in many 'land races', or plant varieties adapted to local tastes and conditions. In a self-pollinating crop such as barley, these land races (many of which are still in existence today) are composed of many different pure-breeding lines, each of which might have a selective advantage under different environmental pressures. In a cross-breeding crop like maize, land races consist of a genetic continuum with a spectrum of traits across the local population.

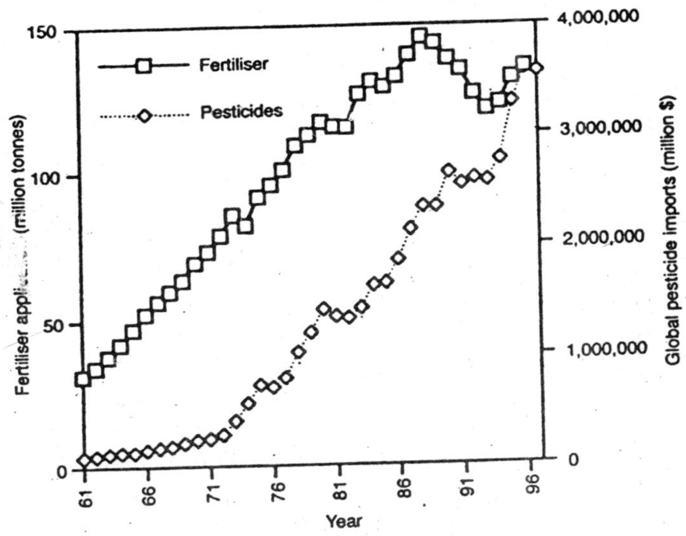

Figure 1.5: ***Annual global fertililzer use and pestic imports over years 1961-97.***

Early farmers must have recognized that like begets like, and this allowed for deliberate selection of positive traits, and for some directed crosses to combine these primitive traits into one plant. Early record show that date palms were deliberately cross-bred some 5000 years ago. Improvements in crops by breeding depend on two factors; eliminating unwanted characteristics and fostering desired characteristics. Desirable traits can only be selected for if they exist in the local gene pool. Trade and travel would have allowed some limited flux in the gene pools of these early crops. With the coming of global expansion by the European 'superpowers' of the seventeenth century, more and more plant species and varieties became available for farmers to use as crops, and plants breeding was widespread by the early eighteenth century. The recognition that spontaneous mutations or 'sports' could be a source of desirable variation also came about at this time. Notable achievements by the early plant breeders include the crossing of two strawberry species, one from North America and one from Chile, to produce the origin of the modern cultivated strawberry.

The rediscovery of Mendel's laws of inheritance allowed progress to be made in the scientific breeding of crop plants. It was recognized that desirable agricultural features are determined by genetic loci that could be passed on the offspring of a plant. The genetic mechanism became understandable and hence more controllable. It became apparent that plants and animals generally have two sets of genes in each cell (the organism are termed diploid), and that the phenotype,. or the characteristics shown by an individual, is a reflection of the expression of these genes. For any given gene, different variants or alleles of that gene exist, which have a dominant or recessive relationship to each other. In a diploid organism, phenotype of the reccesive genes can only be expressed if both copies of the pair of genes in the cell are recessive (the homozygous state), whereas dominant genes give a phenotypic expression even in the presence of one copy of the recessive allele (the heterozygous state).

MODERN PLANT BREEDING

The object of plant breeding is to improve the quality of the crop. Quality is a subjective term, but might include such traits as yield, flavour, disease or pest resistance, and uniformity. These traits are encoded in the genes that are passed on to the offspring by the parents. The mechanisms of plant breeding are to select for desired attributed within a population, or to introduce traits into that population. Introduced traits might arise within the same species naturally, or may be mutant alleles (spontaneous or induced) of a gene, or may be carried on genes introduced from a species that does not normally breed with the crop species.

Approaches to plant breeding depend on whether the crop is self-pollinating (selfing or in-breeding) or cross-pollinating (or out-breeding). Self-pollinating crops such as wheat, oats, rice and barley have physiological and anatomical mechanisms that ensure that individual flowers and primarily self-fertilizing. As a consequence of this, self-pollinating plant population are composed of individuals, each homozygous for the vast majority, if not all, of the genetic loci, and the progeny of such plants will be identical to the parent, or 'breed true'. In contrast, cross-pollinating plants such as maize exhibit mechanisms to encourage pollen transfer between plants. Cross-pollinating plant populations are composed of individuals with a great degree of genetic heterzygosity. Sexual reproduction by a plant carrying heterozygous genes will result in segregation of alleles in the progeny and consequently a phenotype segregation; the offspring will be variable in character.

BREEDING STRATEGIES FOR IN-BREEDING CROPS

Selection within in-breeding crops may use singly plant or mass selection. An existing mixed population of plants composed of many individuals, each being homozygous, but for differing patterns of alleles at each genetic locus, is subjected to selection for the criteria determined by the breeder. In single plant selection, a large number of individual plants are selected out of the variable population, and compared to each other in subsequent sowings. In mass selection, inferior plants are simply culled.

A technique termed pedigree breeding is the most common method of breeding selfing crops. Pure breeding lines of documented and complementary performance are selected and crossed. The next generation, the F1, will be heterozygous for those loci in which the parents differed. The f1 is self-fertilized and singly plant selection takes place in the F2 and subsequent generations. By the F6, after continued self-fertilization, most lines will be homozygous once more, but each line will have a different pattern of alleles at each variable locus. Other traits, not presents in the two original parental lines, can be introduced by crossing them in the F1 stage of the first cross.

Frequently, an established variety (A) may require improvements by the introduction of only one or a few traits from another variety (B). This can be achieved by back-crossing; making a cross between the two parents (A × B), then back-crossing the F1 to parent A. Selection for the desired character is carried out in the F2, or one generation later (depending on whether the trait is dominant or recessive), and at each subsequent stage before another round of backcrossing. Eventually the progeny of the cross will be homozygous for all the alleles in the recurrent parent A and will contain only the desired trait from B.

BREEDING STRATEGIES IN OUT-BREEDING CROPS

Populations of out-breeding crops share a common gene pool and breeding strategies for these crops are designed to enhance the frequency of favourable genes, and reduce the frequency of disadvantageous genes within that pool. Single plant selection followed by enforced self-fertilization to ensure homozygosity for favourable traits is often accompanied by a general loss of vigour, termed in-breeding depression. This is attributed to the accumulation in the homozygous state of deleterious recessive genes, normally masked in the heterozygous state.

Mass selection has been a very effective strategy in improving traits such as sugar content of sugar beet, or for oil and protein content in maize. This can be refined by line breeding, which is mass selection followed by single plant selection and subsequent mixing of these lines. The back-cross technique can also be used with out-breeding crops, except that here a small population of plants are used as the recurrent parent.

Another breeding strategy exploits the phenomenon of hybrid vigour. The in-breeding depression caused by homozygosity in out-breeding crops has a corollary, termed heterosis or hybrid vigour. When two in-bred lines are crossed, the F1 generation frequently out-performs the parents. This breeding strategy is often used for commercial crops. In the first instance, homozygous in-bred lines are developed and deleterious traits (infertility, dwarfness, defective seeds, etc) are removed from the population. In this way, undesirable genes are removed from the line. Continuous selection within lines for normal plants with desirable traits results in homozygous in-bred lines. Crossing two complementary in-bred lines with good general combining ability will give a uniform F1 generation with the attendant advantages of heterosis. Uniformity is an important advantage of the F1 because an out-breeding crop is normally heterogeneous in yield and quality. The skill of the breeder is in ascertaining which two in-bred lines will give an advantageous F1. To aid in the commercial scale production of F1 seed, in bred lines carrying maly sterile genes in one of the parents are used, eliminating the need for hand emasculation. Restorer genes (that restore fertility) in the partner of the cross ensure that the progeny is fertile and will produce the crop. The progeny of the F1 will segregate and revert to a heterogeneous crop, hence the farmer is dependent of the seed company for next year's crop of the same quality. No doubt it would be possible to select similar advantages from an open-pollinated line of maize but for obvious financial reasons, maize breeders have chosen not to do so.

GENETIC DIVERSITY

Whatever the breeding strategy,an important prerequisite for plant breeding is genetic diversity. Without different genes and alleles of genes, there will be no chance for improvement of our crops. To an extent, genetic diversity exists within the crop varieties that are currently in the fields, and can be induced by application of mutagens. But it is critical that we retain the land races and wild varieties of crop plants that are to be found throughout the world, as a source of genetic variation for the crops of tomorrow.

CONCLUDING REMARKS

Biotechnology is a difficult term to define since the harnessing of any biological process to human aims and desires could justifiably be called biotechnology. However, the revolution in our understanding of the molecular mechanisms underlying the processes of life, in particular our understanding of DNA, the prime genetic material, has resulted in the ability to manipulate those mechanism to our requirements. This new-found knowledge and ability is loosely termed biotechnology.

There are two main applications of biotechnology to cereals. The first is as an aid to conventional breeding programmes, as outlined above. Physiological or morphological traits are governed by genes carried on chromosomes. The ability to monitor the presence or absence of such genes in plants (even if those genes are in a recessive state or are not otherwise identifiable through the phenotype) is a great aid to plant breeders. This is done through the use of molecular markers, characteristics DNA sequence or fragments that are closely linked to the gene or genes in question. Molecular biological methods allow the monitoring of such markers in many independent individuals, for example those arising from a cross between two cereal varieties. This is a great aid to the selection process.

The second major application of biotechnology is in the ability to transfer genes between different organisms. This means that specific genes can be added to a crop variety in one step, avoiding all the back-crossing that is normally required, providing a major saving of time and effort. Furthermore, those genes that are added need not come from a species that is sexually compatible with the crop in question. Conventional breeding is of course limited to the introduction of genes from plants of the same species or very near relatives. By employing the science of genetic engineering, it is possible to bring into a crop plant different genes from other plants or even bacteria, fungi or animals. Genes are, simplistically, made up of two parts; the coding region which determines what the gene product is (for example an enzyme like α-amylase, or a seed storage protein like hordein), and the promoter, a set of instructions specifying where, when and to what degree a gene is expressed. Coding regions and promoters from different genes can be spliced together in the laboratory to provide genes with new and useful properties (recombinant DNA). For example, if it were desirable for a heat-stable starch degrading enzyme from a fungus to be expressed during barley germination the fungal gene could be attached to the promoter of a barley gene that is normally expressed during germination. These foreign or recombinant genes can then be introduced back into crop plants through the technique of plant genetic transformation. The introduced genes integrate into the plant genome and will be passed on to the offspring in the normal way.

These new approaches to plant breeding are set to revolutionize cereal technology. Already we are seeing the production of crops with properties unimaginable by conventional breeding techniques. We can anticipate cereal crops with improved yields and qualities, and novel, enhanced or optimized properties.

2

Product Development

During the brief history of modern cereal biotechnology, the development of innovative products has captured a great deal of academic interest and corporate attention, with such innovation being viewed as a highly visible source of competitive advantage. However, while there has been extensive basic research on the problems of characterizing genes with specific agronomic effects, very little has been written about the problems associated with products commercialization in cereal biotechnology. The increasing complexity of transgenic plant products, especially those containing metabolically engineered output traits, will raise the cost of launching the next generation of products in cereal biotechnology. Then, more than ever, firms involved in cereal biotechnology will have to pursue two seemingly incompatible goals: increasing product differentiation while reducing manufacturing costs and development times in an effort to commercialize effectively their products.

Developement in Cereal Production

Product development focuses primarily on designing and testing prototypes of the product. Process development can be described as a system for creating and refining an organization's capability to manufacture a product or set of products commercially. While there is a large body of literature on the competitive advantage of efficient process development in mature industries, such as buk chemicals, and an increasing focus on 'design for manufacturability' in assembly products such as automobile production, little has been published on the competitive role of effective process development in agricultural biotechnology.

Within cereal biotechnology, decreasing process development lead times should result in accelearted product development lead times, especially where innovations in process design are implemented on the road toward transgenic product launch. Rapid and effective ramp-up is essential for lowering the need for capital investments in breeding capacity, lowering current production costs, generating sales revenue and recovering development costs. Some of the resulting reduction in manufacturing costs can be passed on the farmers and other consumers to increase end-user acceptance of new products. Finally, innovative process design technologies can extend the proprietary postition of a specific cereal biotechnology if would-be imitators or competitors are unable to determine how to produce the transgenic product either at competitive costs or at sufficient quality levels.

In this chapter I would like to describe how innovative process development, leading to rapid product commercialization, can generate a competitive edge in cereal biotechnology. A review of the overall product R & D cycle for developing transgenic cereals, using transgenic corn commercialization in the USA as an example, should help illuminate how and where various process development activities fit into the broader product development efforts within cereal biotechnology. The product R & D cycle for transgenic plants can be divided into a number of stages that include: discovery, small-scale greenhouse efficacy screening and evaluation, the generation of intellectual property protection and regulatory approval, large-scale field evaluation and breeding and product release. The timing of these various stages for a hypothetical transgenic corn product is outlined in the project management chart shown in.

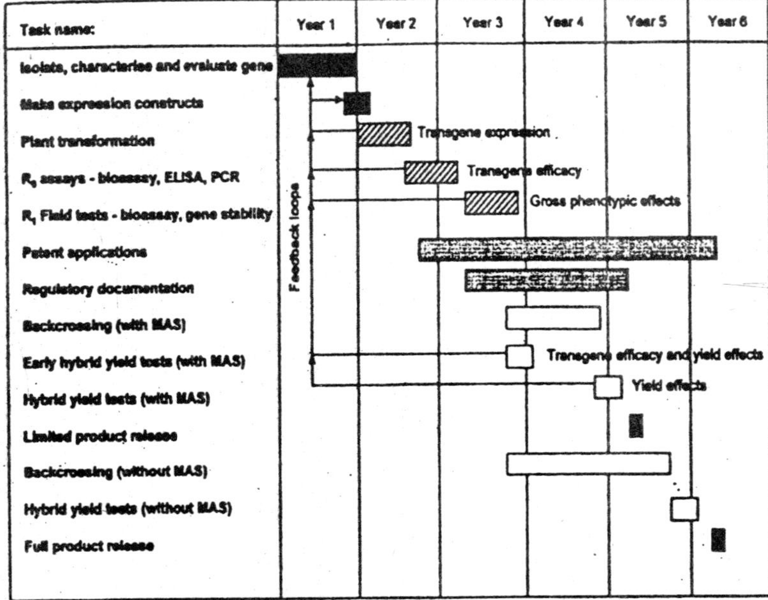

Figure 2.1: ***Project management chart outlining the timing of the various development stages for a hypothetical transgenic corn product. The different steps in the product development life cycle are grouped as discovery (dark boxes), early generation, small-scale traits evaluation (striped boxes), intellectual property and regulatory filings (stippled boxes), lated generation, large-scale field evaluation and trait breeding (light boxes) and product release (grey boxes). Important feedback loops that should exist between the various steps are shown. Approximate time lines for the development of a hypothetical corn product are indicated. MAS = Marker-Aided Selection. Adapted from McElroy.[3]***

Finally, towards the end of the chapter, I will describe the development of transgenic corn products expressing, B.t. transgenes. This example should illustrate how the contribution of each step in the process development cycle leads to the final commercialization of this particular product.

COMMERCIAL TARGETS

One of the major goals of discovery research in cereal biotechnology is to identify genes that safely and effectively generate commercial opportunities in agriculture. The challenge for product designers in cereal biotechnology firms is to close the gap between what product can be offered to the customer, i.e. technology push, and what the customer really wants, i.e. market pull. Technology push is a result of novel product of novel product opportunities arising from, for example, advances in genetic engineering, genomics, information management systems and grain characterization technologies, while market pull comes from increasingly sophisticated end-user and consumer demand for improved processes and products. The products traits themselves can be broadly divided between a first generation of relatively simple input traits, some of which are already on the market, and a second generation of increasingly complex output traits, some of which are about to come onto the market or are in the product pipeline.

Input traits include genes for pest and disease tolerance, such as B.t. genes, α-amylase inhibitors, viral coat proteins, and viral replicase genes; herbicide resistance, including those derived from mutation screening, e.g. resistance to sethoxydim (Poast®) and transgenic approaches, e.g. resistance to glyphosate (Roundup®) and phophinothricin (Liberty®); yield stability, including drought tolerance genes, such as barley *HVAI;* and male sterility genes for the generation of hybrid seed and for intellectual property protection. Most of these input trait products generate revenue by lowering the costs., financial and/or environmental, of plant production.

Output traits include genes for commercially valuable oils, proteins and starches; fatty acid modifications; modification of seed storage protein and amino acid profile; and the manipulation of carbon-partitoning for novel starch production. Effort to minimize financial risk in agricultural have encouraged vertical integration, contrast growing and end-product orientation with an associated focus on the measurement and categoriszation of seed quality characteristics. Most of these output trait products generate revenue by altering the specific measurable properties of the final product to meet these increasingly sophisticated consumer demands.

PROBLEMS IN CEREAL BIOTECHNOLOGY

There are high investment costs associated with the long lead time (6-10 years) for cereal biotechnology products to reach the marketplace and start generating revenue, and this is especially true for the output traits, such as those involved in the modification of grain quality. For such output products the impact on the final agronomic performance of the transgene, and/or the effects of the plant genomic DNA flanking the transgene, might not be fully evaluated until well (3-6 years) into the breeding process. There are a number of other risks associated with cereal biotechnology products that need to be borne in mind during the initial design phase. There can be uncertain profitability associated with cereal biotechnology products that need to be borne in mind during the initial design phase. There can be uncertain profitability with some of the 'technology push' concepts at the onset of product development. There may be intellectual property issues limiting a company's freedom to operate with key technologies or encouraging the rapid development of potentially superior technologies by their competitors. Finally, there might be uncertainty associated with regulatory and consumer acceptance issues inhibiting trade and domestic investment, for example, the export of processed grain derived from transgenic plants to the EU.

New products coming out of discovery research do not necessarily ensure lasting profits. Returns from innovation can be competed away unless isolating mechanisms are in place to inhibit imitators. As the mean development time to launch the next generation of output products increases, the time lag between gaining regulatory approval and patent expiration will inevitably shrink. Although patent protection last a number of years following the date of initial filing, some portion of this period will be lost because of the time required to 'fine tune' the technology, conduct efficacy trial, secure regulatory approval and produce the finished transgenic product in sufficient sales volumes. Instead of conducting their own lengthy and costly regulatory procedures, the producers of generic (off-patent) cereal biotechnology products, which could start coming onto the market in the next decade, might need only to show that their versions of the products is chemically and biologically equivalent to the original patented version. Therefore, for off-patent products, firm-specific differences in their process knowledge base will become an increasingly valuable source of competitive advantage.

COMMERCIAL TRAITS

The development of a cereal biotechnology process plan begins with the design of those analytical methods, screens and other assays that will be used to detect the expression of the desired gene in early generation plants. However, the evolution of analytical methods should continue throughout the project, with later development emphasizing methods applicable to production quality control during breeding and foundation seed production. Analytical techniques play a critical role in evaluating process R & D experiments throughout the product development life cycle by helping to generate a deeper understanding about underlying cause-and-effect relationships in the recombinant trait's efficacy. Particularly novel products will require the invention of novel analytical techniques, tools and procedures. The ability to recognize the need to develop novel analytical techniques in a timely manner is an important cabability within any product commercialization group.

TABLE 2.1 : MOLECULAR TECHNOLOGIES COMMONLY USED IN TRANSGENIC PLANT ANALYSIS

Technology	Application	Generation*
Transgene-specific PCR	High throughout transformant screen, high throughout transgene segregation, regulatory approval, e.g. *Amp*$^{-*}$.	R_0/R_1 R_1/BC_1 BC_1
Event-specific PCR	High throughout QC in backcrossing.	BC_n
Quantitative PCR	Copy number determination, zygosity determination.	R_0/R_1 S_n
Southern analysis	Transgene complexity/copy #/fingerprint, multi-locus transgene segregation, regulatory approval, e.g. *Amp*$^-$	R_1, BC_n R_1 BC_n
RT-PCR	Semi-quantitative transgene expression. regulatory approval, e.g. *Amp*$^-$	R_1 BC_n
Northern analysis	Transgene exprcssion, transcript characterisation, e.g. size, regulatory approval, e.g. *Amp*$^-$	R_1 R_1 BC_n
Western analysis	Transgene expression, transgene product processing, quantity,	R_0/R_1 R_0/R_1

	regulatory approval, e.g. *Amp*$^-$	BC_n
Quantitative ELISA	High throughout transgene expression, high throughput transgene product quant.	R_1 R_1/BC_n
Qualitative ELISA (DipStick)	High throughput transgene expression.	BC_n/S_n
Quantitative analytical	Transgene effect, e.g. lysine by HPLC	R_1–S_n
Phenotype assay	High throughput transgene expression, e.g. herbicide resistance	R_0–S_n

** *Amp*$^-$ = free of any bacterial selectable marker gene, such as that encoding resistance to ampicilling.

Screening methods generally fall into two major types, selective screens or analytical screens. The development of selective screening systems, e.g. for herbicide tolerance, may begin in tissue culture, followed by initial greenhouse trials and experimental fields where promising transgenic lines are evaluated in randomised, small-scale plots. The development of such selective screening systems has been important in the evaluation of input trait efficacy, where many of the products can be evaluated in early generations in phenotypes that are readily transformed but are generally not commercial varieties. Analytical screening systems, e.g. for transgene presence, event verification, oil content, amino acid composition and other grain quality characteristics, may be followed by a subsequent functional assay if such an assessment is critical for the initial identification of efficacious lines. For those outputs traits affecting grain quality characteristics this may involve small-scale compostitional analysis followed by large-scale animal feeding trials or large-scale processing experiments. Many grain quality characteristics can be fully evaluated only in the finished inbred line either to be sold directly or used in hybrid seed production. The generation of such finished lines in both time consuming and expensive.

During the early generation evaluation phase of the product development cycle, the company's goal is to collect sufficient data on the performance of the candidate gene to warrant the much more expensive step of large- scale field evaluation and entry into the breeding program. At this early stage gene candidates can be abandoned and/or targeted for redesign as a result of either a lack of deomstrated efficacy, gross agronomic abnormalities or phytotoxicity.

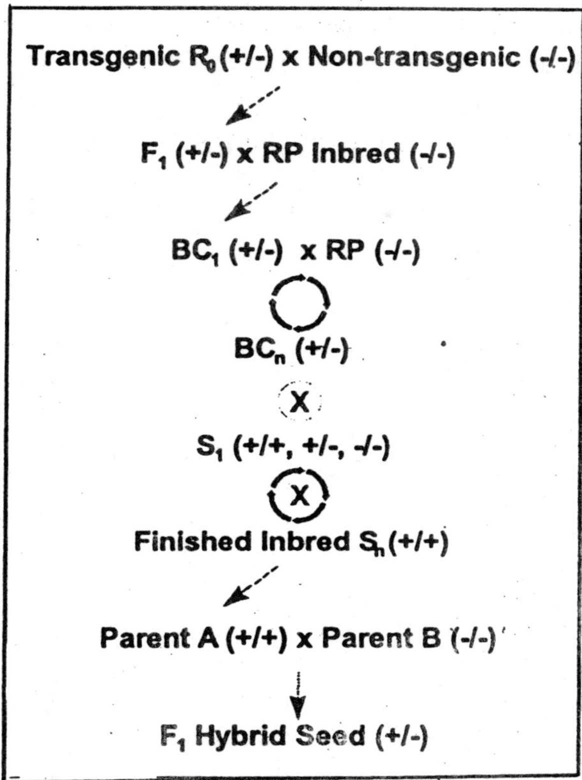

Figure 2.2: ***Outline of the breeding process for a transgenic corn product. The genotypes of the various generations are indicated as either present (+) or absent (–) for the transgenic trait. Abbreviations used include: RP, recurrent parent inbred line; BC, generation resulting from a backcross; X, selfcross; S, generation resulting form a selfcross. A rolling circle with arrows indicated and iterative process.***

The precise principles governing the expression of recombinant traits in transgenic plants are poorly understood. As a result, problem solving in cereal biotechnology relies more heavily on physical experimentation than on conceptual modelling. Thus, while early greenhouse observation are critical the real uncertainty lies in predicting a recombinant trait's agronomic performance in elite commercial germplasm. This is especially true for output traits the efficacy of which can be significantly influenced by genotype effects. In this environment one tends to fine that 'learning by doing' is the only available process development strategy.

TABLE 2.2 : CORN BACKCROSS CONVERSION–EFFECTS OF MARKER-AIDED SELECTION

Period	Activity	Product*	Trad. Mean % RP*	Marker-aided Mean % RP*
Yr1	Backcross (BC) R_0 transgenics to elite inbreds (recurrent parents, RP)	F1	50.00	50.00
	select transgene - BC to RP	BC1	75.00	82.99
	select transgene - BC to RP	BC2	87.50	97.89
Yr2	Select transgene - BC to RP	BC3	93.75	~100.00
	select transgene - BC to RP	BC4	96.87	-
	select transgene - BC to RP	BC5	98.44	-
Yr3	select transgene - BC to RP	BC6	99.22	-
	select transgene - BC to RP	BC7	99.61	-
	self	S1	99.61	~100.00
Yr4	Self	S2	99.61	~100.00
	Identify transgene expressing line phenotypically similar to RP	S3	99.61	~100.00
	Testcross	S3 TC	49.81	~50.00

* $E[p(n)] = [1 - (1/2)^{n+1}] \times 100\%$, with associated variance and S.D. a function of genetic map length and no. of chromosomes.

* Inentify plants carrying the transgene prior to tissue sampling for efficient selection.

MOLECULAR BREEDING OF TRANSGENIC PLANTS

The aim of the breeding program is to develop high-yielding competitive varieties free of any yield drag or other agronomic problems that might be associated with either the novel gene's expression or its initial genetic background. This breeding process might take 5-9 years to complete in corn, depending upon the number of backcrosses involved. Once early (R_0/R_1) generation transgenic lines have been characterized they will be entered into a breeding program with a number of quality inbred lines to do initial assessments of genotype and environmental effects of the performance and genetic stability of the new trait (1-2 years). The efficacy of the novel trait must be assessed through several seasons in multiple locations to ensure that performance standards are consistently met in a number of unique environments. Those transgenes and transgenic events that make it through this initial field evaluation will eventually enter into a variety development program, involving crossing into other high-quality line (3-5 years). Towards the end of varietal development, seed production begins to generate enough seed stocks for commercial release through normal foundation seed multiplication channels (1-2 years). Individuals transgenic 'events' are eliminated at each step in the breeding program as they fail to meet performance benchmarks, so the initial evaluation process must start with a large number of high-quality transgenic lines.

Shortening Breeding Programs

A number of processes can be used to shorten transgenic breeding programs. The development of transformation protocols for cultivars containing a genetic background common to that of elite commercial varieties will reduce the number of backcross generations needed to produce a fully converted transgenic inbred line. Winter nursery facilities can be used for year-long production of multiple product generations. Double-haploid plant production can be used to produce homozygous transgenic lines for rapid evaluation of transgene copy number effects, yield drag associated with individual insertion events or combining ability for stacked products containing more than one transgene. Finally, marker-aided selection can be employed to facilitate greatly the selection of progeny for use in subsequent backcrossing to recurrent parent lines.

Marker-Assisted Selection for Transgenic Plants

Valuable alleles and/or transgenes can be tracked in breeding populations using genetically linked molecular markers. A number of marker systems are available for this process including: restriction fragment length polymorphisms (RFLPs); simple sequence repeats (SSRs); amplified fragment length polymorphisms (AFLPs); and single nucleotide polymorphism (SNPs). Marker-aided selection (MAS) is used to select for genetic similarity to the targeted elite inbred (recurrent parent) within early backcross generations. MAS facilitates the elimination of a number of generations from the backcross conversion process and might save 1-2 years in transgenic product development. MAS also allows one to select for less linkage drag associated with the 'non-elite' genomic DNA flanking the transgene, increasing the probability of obtaining an acceptable conversion. It is advisable to use a sufficient number of markers to sample the genome adequately, with the actual number depending upon the genetic relationship of the transgenic line to the recurrent parent, the backcrossing strategy and the breeder's experience. Uniformly distributed markers are most effective in backcross conversion and minimizing linkage drag than an equal number of markers per chromosome or randomly assigned markers. Using MAS to accelerate backcross conversion is critical to the introduction of value-added traits in elite genetic backgrounds, the efficient management of seed inventories and the timely commercial launch of new transgenic products.

MOLECULAR QUALITY CONTROL FOR TRANSGENIC PLANTS

Transgenic plant lines containing independent transgenes and independent transgenic insertion events are backcrossed and evaluated at a number of different breeding locations. This is done, in part, to evaluated transgene efficacy under a number of different environmental conditions. Within this breeding and evaluation process there exists the potential for line and/or gene misidentification. It is important that events should not be mixed up in the breeding process since only one event might be deregulated for eventual sale. Finally, at the end of the breeding process, the customer must be provided with an estimate of the quality of the transgenic product that they are purchasing, where quality is represented by the percentage of 'off types' they can expect in their field of transgenic plants.

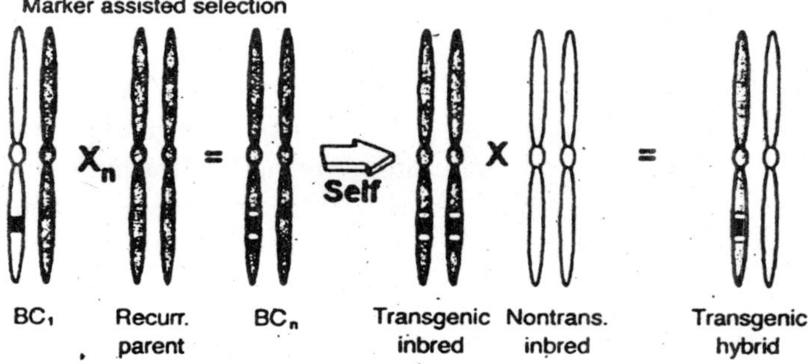

Figure 2.3: ***Overview of breeding a transgenic corn product. (a) Plant transformation and the production of the first backcross generation (BC_1) following crossing fo the R_0 plants with a recurrent parent inbred line. The transgene event ■ is shown integrated into the hemizygous R_0 plant chromosome ■ with the homologous chromosomes for the recurrent paretn shown in* (b) Production of the transgenic hybrids. Following several rounds of backcrossing and selection with the recurrent parent a final backcross generation (BC_n) is produced. This hemizygous BC_n generation is selfed, homozygous transgenic lines are identified and the homozygous material bulked up in foundation seed nurseries. The homozygous transgenic lines are then crossed with other elite inbred lines to generate the transgenic hybrid seed products that will be sole. Adapted from McElroy.***

In breeding programs transgenic lines are often identified by screening for the presence of a more readily detectable selectable marker gene (e.g. herbicide resistance) that was co-introduced with the transgene of interest. Insertion event-specific transgene silencing, recombination-mediated transgene deletion and/or pollen 'contamination' may result in loss of expression for the non-selected transgene of interest. Without a quality control system in place, generations of selection for the readily detectable marker gene may pass before the presence/efficacy of the transgene of interest is evaluated. Therefore, independent transgenes and transgenes insertion events must be tracked and verified thoughout the breeding and evaluation process to ensure that transgenes are not 'lost' or transgenic lines misidentified. This gene tracking process has been dependent upon the development of a number of analytical tools, including rapid transgene and insertion event-specific assays. The gene tracking process has also been facilitated by developments in the automation of DNA isolation/analysis and in date processing.

The recent evolution of quality control system in cereal bio-technology is a good example of prior production experience or 'learning by doing', generating data that can contribute to future process development projects, thus allowing for 'learning' to take place 'before doing' subsequent process biotechnology, manufac-turing performance should improve with cumulative experiecne. Cumulative production experience generates data for systematic problem identification and problem solving, leading to improve-ments in product design, equipment modifications and worker training. However, for such 'learning by doing' to be retained and utilized, there need to be strong feedback loops from the produc-tion process (in this case the breeders) to process R & D teams .

INTELLECTUAL PROPERTY

Freedom to operate (FTO) under any dominating patents must be secured in order to commercialize a product. Transgenic plants often utilize several patented technologies, each owned by different parties, e.g. trasnformation method, selectable marker gene, specific genes, and gene expression elements. Therefore, companies must commit substantial resources to develop either inhouse, or acquire rights to utilize, all of the technologies used in the development of their commercial products. These issues should ideally be addressed before the product enters the breeding program for large-scale field evaluation.

In order to obtain a US patent, the technology must be novel (no prior art), non-obvious (no analogous prior art) to one ordinarily 'skilled in the art', and have some defined utility. The conception and reduction to practice of the claimed invention (the actual invention and any modifications to the invention that retain the same activity that are taught in the patent application) should be described in such a way that anyone in the field could make and use the invention. From a product development point of view in cereal biotechnology patents may be of essentially two major types. A product patent covers compositions of matter such as genes and gene products, monoclonal antibodies, engineered cells, and transgenic plants. A process patent covers methods such as those for gene cloning, transformation, gene silencing, gene delivery, and molecular breeding.

There are a number of problems associated with patents with respect to process development. First, there is usually a long delay between and initial patent application and the issuing of the patent. In the US, the contents of a patent application are not released until the patent is actually issued. Second, there are often a number of overlapping claims between related patents and there is often inconsistency in the breadth of coverage granted between different countries. Third, there can be significant resource requirement to file the patents and, especially is cereal biotechnology, to enforce the patent. Finally, overly broad patent protection is normally not granted at the present time. Such broad patents are generally not written in such a way as actually to teach anyone skilled in the art how to repeat all of the application's claims and the granting of such broad patent protection is also believed to inhibit subsequent innovation.

All of these patent issues, while problematic from the biotechnology firm's point of view, have been a boon to patent litigation lawyers. Thankfully there are a number of solutions to these patent problems that should be investigated during the product's development. Licensing and cross-licensing of technology between firms can occur in order to exchange rights to patents critical to product commercialization. Mergers and acquisitions,. either partial or complete, can be orchestrated in order to gain patent rights for critical technological component(s) in the commercial product. Finally, firms can engage in technology development to supersede a competitor's patents.

REGULATORY ISSUES

There is a widespread belief that standard plant-breeding procedures are generally not adequate to identify all the potential problems associated with transgenic plants. It is argued that, as a consequence of the diversity of available transgenes, the environmental consequences of genetic modifications cannot reasonably be anticipated. Therefore, countries have established their own agencies to oversee the introduction of new transgenic plant varieties into their environments and the importance of novel plant products, particularly transgenics, into their marketplaces. In the establishment of these national agencies, most government have striven to balance the need to develop oversight mechanisms that address potentially serious risks without generating a regulatory burden which creates disincentives to innovation and diminishes the availability of important new technique and products.

Securing government regulatory approval for product introduction and addressing costomer acceptance issues are critical aspects of product development and commercialization, e.g. for key export markets. In addition to gaining approval of the transgenic 'event' itself, e.g. for herbicide resistant crops, the registration and deregulation of any associated chemicals used with the transgenic varieties may also be required. In most countries there are two kinds of regulation that govern research and development of transgenic plants. Rules for 'contained use' govern genetic modification in the laboratory, concentrating mainly on worker health and safety issues. In the US the rules for 'field release' focus on environmental risk assessment appropriate to the nature and final use of the transgenic plant, with each release initially considered on a case-by-case basis in order to build up experience with particular crop/transgene combinations.

A number of factors are borne in mind during the risk assessment process asociated with transgenic plant deregulation. These factors include: the function of the gene in the donor organism; the effect of the transgene on the phenotype of the transgenic plant; the risk of the transgenic to animals and humans, e.g. evidence of toxicity and/or allergenicity; the ability of the transgenic to colonize and persist in agricultural habitats (weediness); and the proximity of the transgenic species to its' centre of origin'. These last two factors address the likelihood and consequence of transgene movement to other weedy or wild relative plants by cross pollination or to other (pathogenic) organisms by horizontal gene transfer. Other risk issues include the impact of the transgenic plant on the evolution of target organism, e.g. insect resistance to B.t.; the behaviour of non-target organism, including minor pest or beneficial insects; the existing ecological relationships within the agricultural system, such as the plant's potential to disrupt existing pollination systems; and the ability of the transgenic to provide a resource for organisms that the pests of other crops within the agricultural ecosystem.

Individual nations generally employ one of two different kinds of regulatory systems, with the data assessed in both being similar. Horizontal regulatory systems, such as those within the EU, are process-based systems that apply to all plants produced by a certain method, e.g. by a particular transformation method. Vertical regulatory systems, such as that in the US, are product-based systems that define the characteristics of modified plants that require them to be regulated. So, except for FDA approval, the US has techniques-based regulatory system which differentiates between 'natural' exotics, genetic improvements by conventional methods, and manipulation by recombinant DNA techniques.

Within the US, transgenic plants are regulated by three government agencies, following a 1986 decision to regulate transgenics under existing statutes. The US Department of

Agriculture (USDA/APHIS) controls permits for inter-state movement and field release of transgenic materials, assesses the pest potential of transgenic plants and determines when transgenics can be field-grown without notification or permits, otherwise known as deregulation. The Food and Drug Administration (FDA) determines that a transgenic plant has been adequately evaluated in accordance with its biotechnology food and feed policy, e.g. for the safety of antibiotic selectable marker genes. Finally, the Environmental Protection Agency (EPA) regulates transgenic plants with pesticidal properties, e.g. B.t. plants, and determines pesticide/herbicide residue tolerance on transgenic plants. In Canada, transgenic plants are regulated by Ag & Agri-Foods Canada and Health Canada. In the Canadian system, products are similarly regulated whether generated by mutation or transformation . This is essentially a risk-based system that focuses on the intrinsic properties of the product, e.g. weediness, invasiveness, or its potential for outcrossing, rather than on the method of production of the variant under evaluation. In Europe, a Novel Foods Regulation went into effect in 1997 to replace completely EU Directive 90/220 that previously covered environmental issues associated with transgenic seeds sales and transgenic grain import. In Japan, transgenic plants are regulated by the Ministry of Agriculture, Forestry and Fisheries and the Ministry of Health and Welfare. To date, there is no international harmonization of regulations to ensure that transgenic plant varieties released in one country will be accepted in another, so antibiotic resistance genes in food products might inhibit international trade in transgenic products. Meanwhile, the Convention on Biodiversity has established a biosafety working group to develop safety standard for international trade in transgenic products.

MARKETING STRATEIGES

In order to recover its substantial R & D investments, the developer of a commercial product might either generate and market the transgenic seed directly or negotiate a royalty with a seed company/companies, e.g. for technologies discovered by universities, government agencies, technology development companies or large agrochemical companies. Because of the previously described learning curve effects in process development, new products might have higher manufacturing costs that older products. Thus the profit margins on a newly launched product might be small or even negative during the first few years of its commercial life.

In transgenic plants, the product is carried in a vegetatively or sexually reproduced form which the customer could potentially propagate following its initial purchase. For hybrid crops, new seed must be purchased each year, therefore the transgene is protected and premiums for the technology can be added to the seed. However, for inbred or vegetatively propagated crops, where transgenic material could be replanted each year, the transgene is not biologically protected; therefore farmers might be obliged to take a licence from the seed company that allows them to buy and grow seed containing the transgene. Alternatively, firms might need to develop biological processes to protect their investment, such as the so called 'terminator' technology.

The marketing strategy for any one particular trait will be influenced by the nature of the product. Inputs traits, such as those conferring herbicide, disease or stress tolerance, while enhance yield or reduce inputs, can return revenue either by increasing the seed company's market share and/or increasing seed sale premiums. For some products, such as herbicide tolerant crops, the agrochemical companies that helped these products can benefits from increased chemical sales, e.g. Roundup-Ready Corn or Liberty-Link® Corn. For output product, such as improved grain quality, there may be either a seed sale premium, an end-product premium for cantrast-grown crops or a mechanism for sharing the added value with the end-user, such as a food-processing company.

A PRACTICAL EXAMPLE

A review of the overall product R & D cycle for developing transgenic B.t. corn, should help illuminate how and where the previously described product and process development activities fit into the broader product commercialization efforts within cereal biotechnology.

TABLE 2.3 : SPECTRUM OF ACTIVITY OF B.T. GENES

B.t. genes	Spectrum of activity	Transgenic product
Cry 1 Ab	European corn borer, southwestern corn borer corn earworm, fall armyworm	Corn
Cry 1 Ac	Tobacco budworm, cotton bollworm	Cotton
	European corn borer, southwestern corn borer, corn earworm, fall armyworm	Cron
Cry9c	European corn borer, southwestern corn borer	Corn
Cry3A	Colorado potato beetles	Potato

B.t. Genes, Proteins and Mode of Action

Bacillus thuringiensis bacteria produce insecticidal proteins, each with their own spectrum of activity. These accumulate as crystals and are thought to aid in bacterial spore germination and vegetative growth following ingestion and gut wall destruction in certain vector organisms. B.t. protein specificity is determined by its interaction with aminopeptidase-N receptor molecules on the luminal side of the insect midgut epithelium, B.t. protein toxicity appears to be determined by the formation of a pore in the cell membrane, which leads to cell lysis and disintegration of the midgut epithelium. The B.t. crystal protein consists of three domains: a helices responsible for pore formation; a receptor binding domain: and a third domain of unknown function. By exchanging corresponding domains between crystal proteins their insecticidal activity can be exchanged, expanded and/or improved.

There are a number of potential advantages associated with the development of transgenic B.t. crops. Their development and deployment will reduce the cost, time and effort spent protecting crops from insects and, relative to many commercially available chemical insecticides, they should contribute to an environmentally friendly crop production system.

Development of Transgenic Crops Expressing B.t. Transgenis

B.t. corn with improved European corn borer (ECB) resistance was first released to farmers in 1996 and a number of competing B.t. corn products are now on the market. However, a number of product design and process development issues had to be addressed in order to develop and commercialise transgenic crops expressing B.t. transgenes.

TABLE 2.4 : B.T. CORN EVENTS REGISTERED WITH THE EPA AND SEED MARKET SHARE ESTIMATES

Company	B.t. genes	Transgene expressions	B.t. events	Trademark	Est. 1998 acress, %
Monsanto	*CryIAb*	Constitutive	MON810	YieldGard*	76
Novartis (NK)	*CryIAb*	BT11	YieldGard*		
Novartis (Ciba)	*CryIAb*	Green tissue	176	Maximizer*	17
Mycogen	*CryIAb*	& pollen	176	NatureGard*	
DEKALB	*CryIAc*	Constitutive	DBT418	Bt-Xtra*	5
AgrEvo	*Cry9c*	Constitutive	PGS351	Starlink*	1

Efficient transformation and selection systems had to be developed for all major crops with the choice of transformation protocol used by any one company being influenced by intellectual property considerations. Efficient transgene expression systems needed to be generated. These included the use of synthetic B.t. genes optimized for plant codon preference, truncated B.t. genes and highly active constitutive promoters for driving high level expression in transgenic plants.

Companies had to establish efficient transgenic plant production systems. Gene transfer can result in multiple transgene integrations and rearrangements of heterologous and/or homologous DNA with associated effects on transgene expression. In many cases a large number of transgenic events needed to be produced in order to identify the few lines with adequate transgene expression levels and agronomic performance. Transgenic plant identification and characterization methods had to be established and validated. Molecular breeding and molecular quality systems had to be put in place to handle transgenic products.

New regulatory systems for transgenic plants had to be established. For example, in the US the EPA considers B.t. crops as pesticides and they need to be registered as such, including toxicological studies on a range of organisms. Companies hoping to sell B.t. products needed to design and implement resistance management (RM) strategies, with grower recommendations and monitoring protocols to ensure that resistant insects did not evolve so that the lifetime of the product could be maintained. Systems had to be developed to integrate these new technologies into current farming practices. Finally, there was, and continues to be, a need to establish acceptance of transgenic crops and crop products by growers, food processors and consumers.

Insect Resistance Management

The success of B.t. crops will depend on whether target pest develop resistance to them. Specific insect resistance management (IRM) measures have been, and are continually being, designed and implemented to minimize the build-up of resistance genes in insect populations that might develop under constant selection pressure from transgenic crops. A number of different IRM strategies have been proposed. In the 'gene stacking strategy' transgenic plants are produced that express a combination of different B.t. genes, or the other insect resistance genes, each with a unique mode of action that should delay the evolution of resistant insect populations. In the 'high dose + refuge strategy' transgenic plants are developed with a *high dose*, or relatively high level of B.t. protein accumulation, to eliminate resistance genes in heterozygous resistant insects that are only slightly less susceptible to B.t. than fully susceptible insects. At the same time a *refuge* of non B.t. plants is provided to dilute resistance development in insect population. In 1998 Monsanto's IRM Strategy for their YieldGard® B.t. corn allowed growers either to plant at least 5%

of their acres with a non-B.t. corn hybrid and not treat the rest with insecticides registered to control ECB or to plant at least 20% of their acres with a non-B.t. corn hybrid and treat all their corn acres with non-B.t. insecticides as needed.

Commercial Goals of Insect Resistant Corn Product

The goal of insect resistant corn is to redirect a substantial portion of the insecticide market from agrochemicals to the seed industry. Almost 25% of all pesticides used in US agriculture are insecticides. In 1996 the insecticide market was estimated to be ~$7 billion. However, for transgenic plants expressing B.t. genes, it should be borne in mind that their initial ECB target is a cryptic feeder that spend most of its time in the stalk and so it is a pest that is inaccessible to most conventional insecticides. Therefore, it is uncertain exactly how much of the conventional insecticide market has been captured by B.t. plant products.

TABLE 2.5 : ADDED VALUE* PER ACRE FROM USING B.T. HYBRID: % YIELD PROTECTION BY PREVENTING ECB DAMAGE

Yield potential (bu/a)	0%	2%	4%	8%	16%
150	–$10.94	–$3.44	$4.06	$19.06	$49.06
200	–$10.94	–$0.94	$9.06	$29.06	$69.06

* Assume B.t. seed premium is $35/unit (80,000 kernels), one unit plants 3.2 acres (25,000 kernels per acre), corn price at $2.50/bu.

The B.t. technology itself is an insurance product, with the return to the farmer being directly proportional to the level of potential yield loss that would have been caused by insect damage in the absence of any protection afforded by the B.t. transgene. The break-even point is estimated to fall between 2-4% potential yield loss, depending upon the yield potential of the crop. During 1997 the actual yield advantage for B.t. corn over non-B.t. corn in the mid-west US was determined to be between $15 and $43 bushels/acre.

From the onset B.t. corn was expected to increase yield by ~5% in ~20% of the 82 million acres of hybrid corn planted in 1998, representing a potential gross added value of ~$350 million to those who developed this transgenic cereal product. B.t. corn sales in 1998 represent an added gross premium of ~$150 million, with the final net added value being dependent upon the costs assocated with seed production, quality control, royalties to third-party technology providers, market share gain, discounting, dealer costs, etc.

TABLE 2.6 : 1997 YIELD ADVANTAGE FOR B.T. VS. NON-B.T. CORN ISOLINES

State	Yield advantge B.t. vs. non-B.t. (bu/A)	Yield advantage B.t. vs. non-b.t. ($/A)
Iowa	21.6	$43.06
Illinois	17.4	$32.56
Nebraska	14.7	$25.81
Indiana	11.9	$18.81
Ohio	6.3	$15.75

* Assume B.t. seed premium is $35/unit (80,000 kernels), one unit plant 3.2 acres (25,000 kernels per acre), corn price at $2.50/bu.

TABLE 2.7 : US B.T. CORN SEED COMPANY MARKET SHARE ESTIMATES,

Company	**Acres, millions**	**Acres, %**	**Revenue $ million**
Pioneer	7.2-7.8	44	75
Novartis	6.0-7.0	38	39
DEKALB	1.2-1.4	7	13
Mycogen	0.3	2	2
Others (13)	1.3-1.5	9	15
Total	16-18	100	150

FUTURE TRENDS

Efforts to minimize financial risks will continue to encourage functional consolidation, vertical integration, contract growing and end-product orientation in modern commodity agriculture. The resulting agribusiness conglomerates should have a number of common enabling features in order for them to take full advantage of future product commercialization opportunities in cereal biotechnology. They should have access to proprietary, elite germplams so that they can piggy-back their traits onto already high-yielding varieties. They should have the ability rapidly to move traits into finished products with an extensive, efficient seed production, distribution and sales system. They should have a significant market presence (sales > 150,000 units/yr) in order to capitalize fully on the investment, direct or indirect, required to obtain the transgenic trait. Finally, they need to have innovative customers who are capable of responding to new product opportunities. From the description of the product R & D cycle in cereal biotechnology, one can also conclude that a major challenge facing firms today is to create development processes that are fast, efficient, and capable of generating high-quality products. While process development can have a significant impact on a product's manufacturing cost, savings can be missed by failure to invest adequate resources in process development early enough in the product life cycle.

It should now be apparent from the foregoing description that development projects have two major outputs, the technology that is implemented in the new process and organizational knowledge that becomes available for future projects. Organization knowledge includes how to manage projects, allocate resources, assign personnel and resolve disputes. Each new project should generate feedback on gaps between expected and desired lead times, uncover critical parameters, and create insights about the process performance that triggers a search for new approaches to managing future development projects. Much of this new knowledge becomes embedded in the firm's organizational routines and procedures. However, high performance in process development requires the ability to learn,and some organizations will be better able to learn from their experiences than others. Learning is rooted in how organizations feed data from manufacturing (breeding, foundations seed and sales) back into its technical knowledge base allowing R & D to anticipate future problems. Organizational or geographical barriers between process R & D and plant breeding can have a significant negative consequence if R & D scientists are incapable of designing processes that work well in the field.

In conclusion, the success of any individual organization will depend upon its ability to do more proficiently than their competitors those things that their products demand and their customers value. As cereal biotechnology matures, and consumers prices eventually stabilize, the ability to develop rapid and efficient process will be an increasingly important source of competitive advantage in the commercialization of products in cereal biotechnology.

3

Biotechnology Value to Cereals

The value of cereal crops can be improved in two main ways. Firstly, the quantity of crop produced can be increased and secondly, the quality of the harvested grain can be enhanced. Linked to the question of quality is also the improvement of safety. These two approaches are not entirely independent. Improvements in the productivity of cereal varieties can be associated with improved quality of the grain.

Removing constraints imposed by biotic stress (the impact of pests and diseases) is an attractive option for improving the productivity of cereal crops. The introduction of genes conferring resistance to major pests and diseases provides an opportunity to improve productivity by removing the losses associated with the specific disease targeted. The introduction of herbicide resistance into cereal crops allows a reduction of losses associated with competition from weeds. Both these approaches may lead to associated improvements in grain quality. Freedom from weeds can reduce or eliminate the problem of weed seed contamination in cereal grain. Removal of disease constraints may result in improved grain quality avoiding the losses in quality associates with the presence of disease organisms in the crop. Diseases often result in reduced grain size and associated quality deterioration.

The quality of cereal grain may be improved either in a nutritional sense or by improving the processing properties of the grain. Most conventional plant breeding has addressed the need for appropriated processing qualities in new cereal varieties. Biotechnology may allow more emphasis to be placed on novel alterations of nutritional quality.

Molecular markers can be used to imporve the efficiency of cereal breeding programs aiming to improve the value of cereal crops. Molecular analysis may also be applied in fingerprinting or identification of cercal genotypes with more immediate potential for improvement of cereals. For example, molecular analysis of genotypes can be used to monitor seed purity and identity prior to planting and to characterize grain lots in trading and processing. The composition of cereal-based foods or products can be monitored to ensure authenticity of labelling. These applications of biotechnology can have an almost immediate impact on the quality and value of cereal production.

Genetic engineering offers the possibility of going beyond these short-term outcomes of biotechnology applications to the generation of more novel cereals with increased value in the longer term. In the sections that follow, the main benefits of genetic engineering are summarized under the following headings:

- productivity
- product quality
- safety.

WEED CONTROL

The control of weeds in cereal crops may have a major influence on grain yields but usually has a much lesser impact on grain quality. This is because the major effect is in early crop growth impacting more on grain number than size. Late weeds are an exception. Generally, contamination of seed crops with weed seeds is likely to be the major quality defect.

Classes of Herbicides and Available Resistance Genes

Resistance genes are available for many different classes of herbicide. The major groups of herbicides and the genes available for conferring resistance to these herbicides are listed in.

The most attractive herbicide resistance genes for introduction into cereals are those that confer resistance to herbicides that are considered safe in the environment. Herbicides with low mammalian toxicity and little or no other environmental problem may be attractive alternatives to the more specific herbicides currently in use. The development of transgenic cereal with resistance to appropriate herbicides may facilitate the reduction in use of less desirable herbicides in agriculture and food production.

Escape of Herbicide Genes of Weeds

A major risk associated with the production of transgenic cereals with herbicide resistance is the possibility that new weeds may result either from escape of the gene into other plants or by the transgenic cereals themselves becoming weeds. The production

TABLE 3.1 : HERBICIDE RESISTANCE GENES

Herbicide	Mode of action
Glyphosate	Inhibits 5-enolpyruvyl shikimate-3-phosphate
(EPSP)[1]	
Sulphonylureas	Inhibits acetolactate synthesis (ALS)[2]
Imidazolinones	Inhibits acetolactate synthesis
Triazalopyrimidines	Inhibits acetoplactate synthesis
2-dichlorophe--noxyactic acid	Auxin action
Phosphinothricin	Inhibits glutamine synthesis
Atrazine	Inhibits electron transport in photosystem II

1 Prevents the synthesis of aromatic amino acids.

2 Prevents the synthesis of leucine, isoleucine and valine.

of transgenic plants with resistance to herbicides that have unique modes of action is highly desirable. Multiple herbicide resistance may arise if genes target biochemical pathways that are associated with the action of several classes of herbicide. The problem of escape of herbicide resistance genes from cereals is likely to be a more serious issue for species that are out-crossing rather than for those that are predominantly or exclusively self-pollinating. In some species such as rice, weedy varieties have developed as a result of current agricultural practices. The introduction of herbicide-resistant rices could lead to the development of a further class of weedy rices based upon their herbicide resistance if appropriate agricultural practices are not adopted in association with the new varieties. This could require production to be limited to specific regions and to include the rotation of herbicides or varieties. Despite these limitations,. herbicide-resistant cereals should provide enormous advantage in the enhancement of cereal productivity.

DISEASE RESISTANCE

Disease Resistance

Diseases may seriously reduce grain quality. Fungal diseases may have a large impact on grain quality (especially grain size) because they may be active on the leaves during grain filling or directly infect the head. Insect pests may reduce yields and grain quality. Post-harvest damage from insects can be a major problems. Mycotoxins resulting from fungal growth on the grain are a serious safety issue for grain from some environments.

Developement of pest- and disease-resistant cereals provides a major opportunity for enhancing cereal productivity. In many environments single disease may be associated with very serious losses in grain yield.Breeding resistant varieties has been a major strategy used in increasing cereal yields. Transgenic cereals with high levels of disease resistance may extend the options available from conventional plant improvement. Resistance to a wide range of biotic factors may be engineered using appropriate genes. Resistance to viruses, bacteria, fungi, nematodes and insects has been reported.

Native Genes for Disease Resistance

Classical cereal breeding has involved the combination of

disease-resistant genes from different sources to produce commercial varieties with effective disease resistance. Molecular markers or direct analysis for the presence of the required gene are now used to improve the efficiency of selection of disease-resistant lines in breeding.

One option for the engineering of cereals with disease resistance is to manipulate or enhance the levels of expression of genes already present in the genome. Specific options include the expression of defence genes using constitutive promoters such that the defence gene product was always produced by the plant regardless of the presence of a specific pathogen. Promoters induced by the disease are also an important option. This approach allows defence gene products to be produced by the plant only in response to attack by the pest.

TABLE 3.2 : SOME NOVEL PEST- AND DISEASE-RESISTANCE GENES OF POTENTIAL VALUE IN CEREALS

Type of Protection	Gene	Source
Virus	Coat protection	Barley yellow dwarf
	Coat protection	Maize chlorotic mottle
		Maize chlorotic dwarf
		Maize chlorotic mosaic
Bacterial and fungi	Chitinase	Various
	Glucanase	Various
Insects	Bt toxin	*Bacillus thuringiensis*

Novel Genes for Disease Resistnace

Novel genes from plants or non-plant sources may provide durable resistant genes for use in cereals. Examples have been described for bacterial, fungi, nematodes and insects. The use of virus-derived sequences for breeding virus-resistant plants has been a notable success in many species.

Environmental Impact of Disease Resistance

Transgenic plants with resistance to pests and diseases may

have a significant impact on the environment. The risks should be similar to those of traditional resistance breeding. However, if the transgenic strategies prove dramatically more effective they could seriously deplete populations of plant pathogens and even lead to the extinction of highly specific pest organisms. Evaluation of these risks is important in successful applications of transgenic technology.

IMPROVED NUTRITIONAL PROPERTIES

Importance of Cereals in Human and Animal Nutrition

Cereals are a very important part of human diets. The three major species, wheat, maize and rice, account for a large proportion of the calories and protein in human diets. The importance of cereals in the food chain is also attributable to the extensive use of cereals in the diets of animals. The major constituents of cereals are the carbohydrates and proteins. Other grain components such as lipids and vitamins may be of great significance in human nutrition because of the large contribution of cereals to the diet. Biotechnoloy provides new options for manipulation of the nutritional properties of cereal grains. The carbohydrates of cereals include the simple sugars. The more complex oligosaccharides such as fructans, storage polysaccharides of the grain (starch) and the cell wall polysaccharides, all of which are of nutritional value. All of these carbohydrate components are potential targets for manipulation in improvement of cereal quality. (for example, sugar beet has been transformed to produce fructans). Benefits that may result include reduced cariogenic bacteria (dental health), lower energy value and stimulation of beneficial bacteria in the colon. The sugar content may also influence the quality of the grain for various products. Fructans may be considered to be important to human nutrition because of their possible role as soluble fibre. Starch, as the major component by weight of the grain, may have a great impact on nutritional quality. Resistant starches (not digested in the gut) may be considered critical in influencing the incidence of certain human diseases, such as heart disease. The cell wall polysaccharides may also be important as either soluble or insoluble fibre, depending on the composition of the polysaccharides in the cereal product. Soluble fibres may reduce the risk of heart disease while insoluble fibres contribute to reduced risk of colonic cancers.

Cereals proteins are not well balanced in amino acids required in a nutritionally balanced diet, and genetic engineering may provide opportunities to improve the balance of essential amino acids in cereal-based diets. The lipids in cereals are generally of limited importance in human nutrition but may be important in animal diets. The manipulation of iron levels in cereals through the introduction of haemoglobin illustrates the potential application of biotechnology to enchancing the nutritional value of cereals.

Animal Foods

The major requirements of animals differ depending on whether they are monogastric or ruminant animals and the potential of biotechnology to improve the nutritional values for these two classes of animals differs. Ruminants have a much greater capacity to digest cereal fibre effectively. Soluble fibre components may also be important. For example, the β-glucans of barley limits the use of this cereal in the diet of chickens because of its adverse impact on nutrient utilization.

Aquaculture Foods

The increasing shortage of seafoods (declining fish stocks in the oceans and increasing human populations) indicates great potential for enhanced use of cereals in aquaculture diets used in fish farming. Cereals provide a very cheap option and if biotechnology can be used to enhance the nutritional value of cereals as a component of aquaculture diets we can expect wide-scale use of cereals for the the production of aquaculture products. Cereals may have an important role as a binder in aquaculture feeds. Improvements of protein and lipid composition by genetic engineering may produce more useful cereal aquaculture feeds.

IMPROVED PROCESSING PROPERTIES

The quality requirement of cereal processors may be complex as indicated for barley in Table 5.3. This table lists a few of the characteristics defined as requirements of a barley for use in malting and brewing. Establishing opportunities for quality improvement requires a knowledge of the processes used to convert cereals into end products.

TABLE 3.3 : BARLEY QUALITY CHARACTERISTICS REQUIRED FOR BREWING

Character	Requirement
Grain colour	Bright, white aleurone
Grain size	Plump grains, 90% above 2.5mm, as little as possible below 2.2 mm
Protein content	Optimum 10.5%-11.5% (dry basis)
β-glucan content	Low - maximum and minimum not defined
Husk	Minimum required for brewing
Dormancy	As little as possible (but on pre-harvest sprouting)
Rate of modification	As fast as possible (steeping and germination maximum 110h)
Malt extract	As high as possible (only too high if protein on husk becomes limiting)
Malt enzymes	As high as possible (minimum requirements for disantatic power and β-glucanase)

Introduction to the Range of Processes

The processing of cerealsinvol ves a wide range of techniques with very differing raw material requirements. Several of the major processes of cereal processing will be described here in an attempt to identify the major opportunities for biotechnology to be applied to improving cereal processing quality.

Milling

The milling of cereals involves processes that are ependent substantially on the anatomical structure of the grain. The potential for singly genes to be manipulated in ways that enhance milling quality may be limited because of the large number of grain characteristics contributing to milling performance. The shape of the grain and adherance of the various outer layers are of great importance. Hardness is also a key attribute in milling. Grain colour is a key quality attribute that is influenced by the milling process. The levels of pigments and the size of particles generated in milling both influence colour. The contamination of endosperm fractions with more highly coloured outer layers such as bran are key determinants of colour.

Baking

Many components of cereals such as wheat are important in baking quality and are obvious candidates for the application of biotechnology. The requirements of modern high-speed plant bakeries are for a very consistent quality. Characteristics such as dough stickiness associated with some alien (non-wheat) sources of disease resistance in wheat are serious problems in these high-capacity facilities.

The major components of wheat all contribute to baking quality and are thus targets for genetic improvements of baking quality. Proteins are essential for the viso-elastic properties of wheat doughs. Starch and cell wall polysaccharides (e.g., pentosans) also influence baking quality. The breakdown of starch by amylases is a key process in baking. The pentosans of the cell wall also have a significant influence on loaf quality.

Malting

Malting is the first step of processing grain for use in brewing and distillation. The malting of cereals, most specifically barley, is a process of germination. The rate of germination and the changes in the composition of the barley during malting are potentially important targets for biotechnology. The breeding of malting barley varieties focuses on several major quality characteristics associated with the malting process.

Enzymes are involved in the breakdown of cell walls during malting. The β-glucanases expressed during germination are essential to ensure that the levels of β-glucan in the malt are low. High malt β-glucan levels contribute to high wort viscosity, poor wort and beer filtration and potential hazes in finished beer.

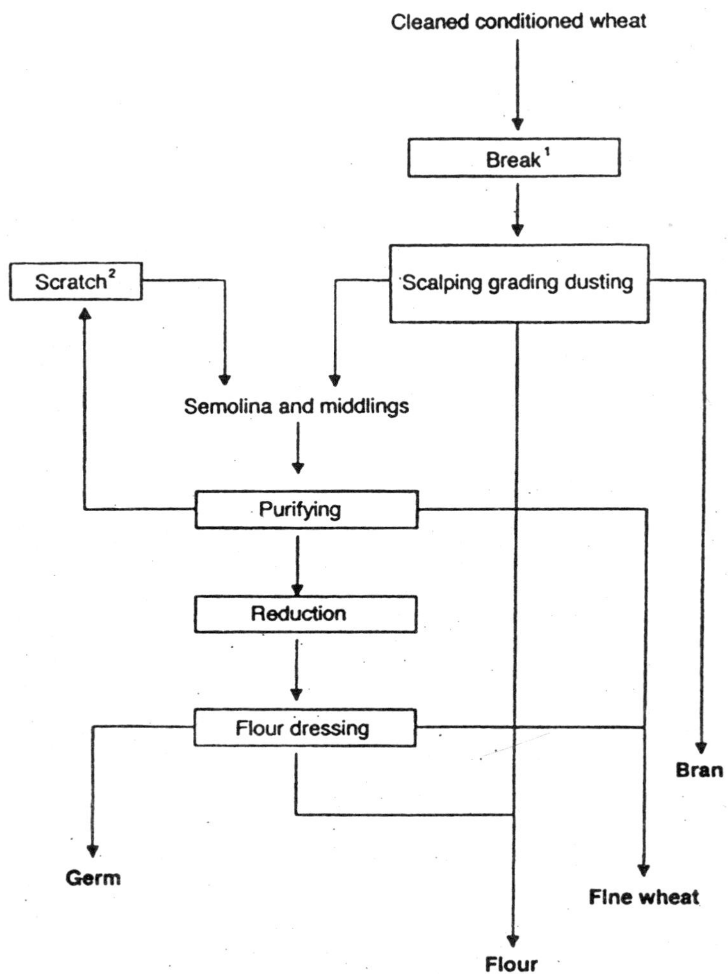

Figure 3.1: ***Schematic of flour milling.***

Brewing

The brewing of beer from malt requires specific malt specifications that are potentially able to be manipulated using biotechnology. Diversification of beer styles and markets are imposing divergent raw material repuirements for the different beer styles.

Sufficient levels of starch degrading enzymes in the malt are necessary to ensure breakdown of starch to fermentable sugars during brewing. The amount of starch (adjunct) added in the form of rice or maize is a key determinant of the level of starch degrading enzyme required in the malt. The relative levels of different starch degrading enzymes are important in determining the nature of the substrate for fermentation and the sugar, alcohol and oligosaccharide content of the beer. The alcohol and residual sugar content (sweetness) are influenced by the levels of fermentable sugars and the non-fermentable oligosaccharides contribute to the taste (mouthfeel).

Distilling

Distillation is a process where the product quality may be less dependent of the raw materials than many processes and may be a less important target for biotechnology application in relation to the cereal raw material. However, the quantity of starch available for fermentation might be enchanced.

Extrusion

Extrusion (production using high temperatures and pressures) of cereal products is an increasingly important process in the production of a wide range of product including snack foods, breakfast cereals and pet foods. The processing properties required are complex but may be enhanced by the application of biotechnology.

Wheat Utilization

Wheat is used for a wide range of products with differing quality requirements. The importance of different wheat grain components and characteristics depends on the ultimate end use products. The protein quality is much more important for products such as breads than for cakes and biscuits. Enhanced levels of desirable high molecular weight glutenins may be desirable in wheat for use in breadmaking. Manipulation of starch synthesis and starch properties may be more important in products such as noodles. Colour may be controlled by a small number of genes and has differing importance. For example, the yellow pigments

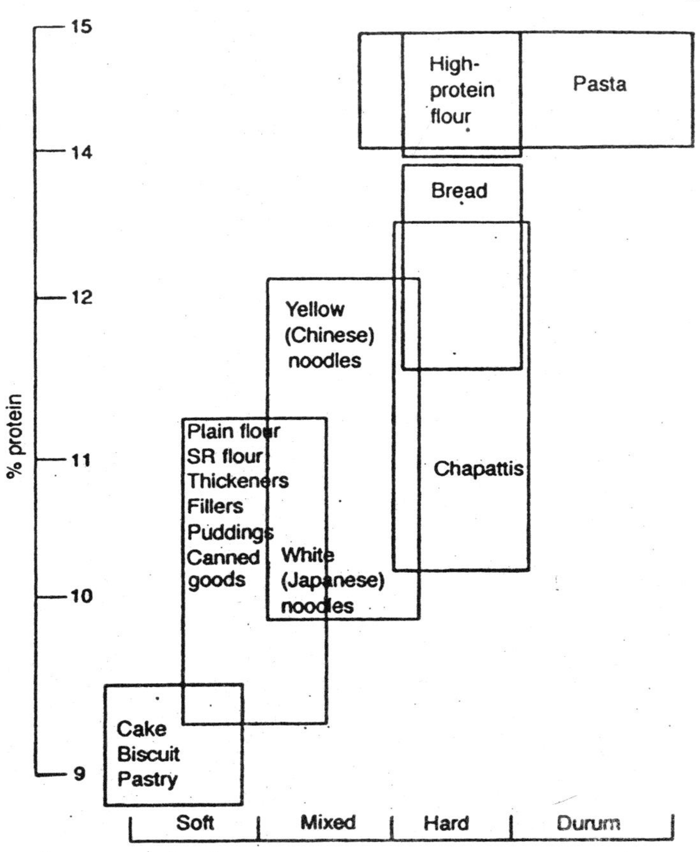

Figure 3.2: ***The broad range of wheat types required for different uses.***

in *durum* wheat are considered highly desirable while some noodle products require very white flour. The imporvement of specific attributes using biotechnology needs to target characters specific for particular end users. Wheat is used to produce the types of food products listed in.

Improvements of the value of wheat for this wide diversity of uses requires the matching of wheat characteristics to specific end product requirements. For example, wheat with different combinations of protein and hardness are better suited to particular end uses (Figure 5.2). However, some combinations of characteristics will not be optimal for any major end use.

Genetic improvement of wheat quality needs to address targets relevant to the end use characteristics of wheat from a particular environment or region. For example,. selection for specific starch metabolism mutants or engineering of improved starch qualities (such as starch-pasting properties) may be important in regions producing noodle wheats while storage protein modification may be more appropriate in regions producing bread wheats.

Beer Production

The production of a specific product such as beer can be the basis for analysis of opportunities for application of biotechnology. The traditional process from barley, to beer is depicted in. Some key attributes of this process may be influenced by the composition of raw (especially the cereals) used and the type of beer to be produced.

Changing composition of barley by genetic engineering might alter performance in any of the many steps in the process of beer production. The production of proanthocyanidin-free barley to reduce haze and improve the shelf life of beer in cold storage has been achieved using mutants and may be more precisely controlled using genetic transformation to block specific steps in the pathways leading to proanthocyanidin formation. One complication of this chane may be that the absence of proanthocyanidin in the wort may result in reduced protein precipitaiton during boiling. The overall result can be a decline in product stability due to increased levels of protein in the beer. This illustrates the need to understand all the interactions during the processing before implementing changes to novel cereal raw grains.

TABLE 3.4 : FOOD PRODUCTS PRODUCED FROM WHEAT

Fermented (leavened) breads	white pan breads	white pan loaves, sandwich (hamburger buns), raisin breads, variety breads
	hearth breads	baguettes, Vienna breads, sourdough breads
	sweet goods	Doughnuts, cinnamon roll, coffee cakes, danis, puff pastries, French brioche
	steambreads	Chinese northern style steambread, Chinese southern style steambread,

		Pan de sol (Phillippines), *Sake-manju* (Japan), *Mushi-manju* (Japan)
Flat breads and crackers	flatbreads	chappati, rotti, naan, paratha, poori, balady, pia, banabri
	tortillas	
	pizza crust	
	muffins	
	crumpets	
	bagels	
	pretezels	
	crackers	soda carckers, cream cracker, water biscuits, graham crackers, sprayed crackers, savoury crackers
Cookies and cakes	pie crusts	
	coodies (biscuits)	
	scones	
	moon cake	
	batters (ice cream cones, pancake and waffles)	
	cakes	
	tempura	
	soup thickeners	
Noodles	alkaline noodles	
	white salted (Udon)	
	soba	
	egg	
	noddles may be fresh, dried, frozen or instant.	
Breakfast foods		
Starch/gluten		
Pasta and durum products	pasta	spaghetti, macaroni, lasagne, fettuccine
	couscous	
	bulgur	
	durum bread	

Molecular markers have been developed for many malting quality attributes and these may accelerate the development of new malting quality barley varieties for use in beer production.

Genetic engineering of barley to improve cell wall breakdown during malting and brewing and lipids contributing to beer flavour have been the subject of active research. Modifications of starch

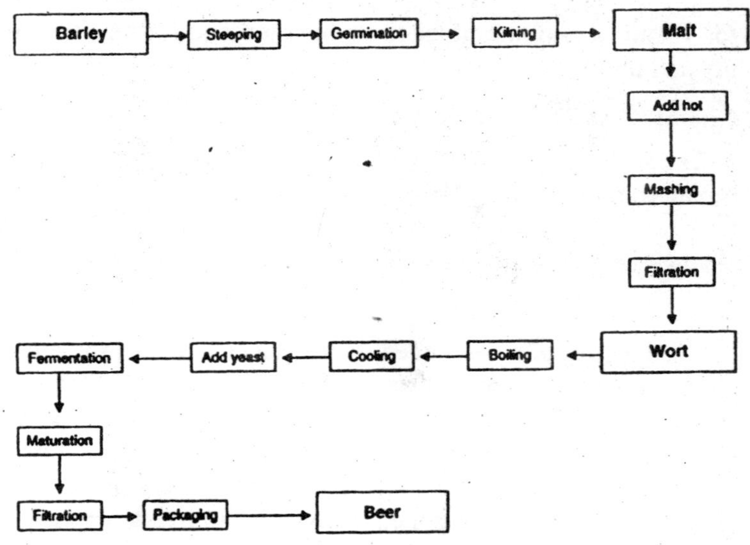

Figure 3.3 : *Traditional beer brewing.*

properties and metabolism and protein composition may also be important. A fundamental limitation is that extract levels are limited by the need to preserve a minimum amount of husk to act as a filter bed in many traditional brewing processes. Grains with extremely high proportions of endosperm necessary for extreme extract levels will have insufficient husk. New filtration technologies could overcome this limitation allowing the development of very high extract barleys. Reliance on enzymes in the malt for starch breakdown during mashing and the issue of yeast nutrition also requires that a minimal protein level be maintained. Genetic engineering could allow a higher proportion of proteins in the malt either to contribute useful enzyme activity or support yeast nitrogen nutrition.

IMPROVED CEREAL QUALITY CONTROL

The value of cereals can be improved by better specification of the identity and composition of cereals in marketing. The major contributions to achieving this type of value addition are quality assurance programs and enhanced tools for analysis of products for chemical and microbial contamination and for genetic identity and purity.

Chemical and Microbial Purity

The strict application of quality assurance principles can ensure cereal safety and value. New technologies allow the use of ELISA tests to establish the level of pesticide residues and DBA-based analysis of microbial contaminants.

Genetic Purity

The identity of a cereal genotype defines many of the quality characteristics such as protein contents and grain size. This can define the processing value and optimal end use of a parcel of cereal grain. Protein content can currently be measured rapidly in the field using near infra-red reflectance (NIR). The development of rapid genotyping methods would find wide application in the cereal industry and allow a relatively complete objective description of samples for commercial valuation.

Distinction of high-value noodle wheats from visually similar wheats of low noodle quality, and distinction of malting and food barleys with similar appearance, are good examples of the areas of potential application of this technology. Genetic purity and level of admixture are also important attributes to assess in commercial trading because of the financial incentive to add low-value grain to parcels of very high value but visually similar genotypes. Rapid DNA extraction from grain is a key technical requirment for successful application of these methods.

4

Molecular Biological Tools in Plant Breeding

PLANT BREEDING

The small-grained cereals considered in this chapter, namely wheat, barley and oats, are all natural inbreeders, e.g. any one plant is more likely to reproduce with itself rather than with a neighbour. This mens that, once a potential plant variety is homozygous, there is no residual variation and it will reproduce itself exactly from one generation to another. Breeding programmes for the small-grained cereals generally follow a pedigree selection scheme with minor variations in detail. At the beginning, breeders cross parents which complement each other for desirable characteristics to produce the F1 generation which will be genetically uniform. The F1 generation naturally self-pollinates to eventually give rise to a population of inbred lines that will contain 2^n different genotypes, where n is the number of loci at which the two parents differ. With differences at just 20 loci, the number of potential different inbred lines in 3^{20} or 1,048,576-a large number that would occupy about a quarter of a hectare at UK commercial sowing rates if each line was represented by just one seed. To be sure of generating one particular gene combination, the population size would need to be larger still and when one considers that a breeder will make several hundred crosses a year it is abundantly clear that plant breeding programmes cannot accommodate all the possible combinations. Most breeding programmes are therefore constrained to several hundred thousand F2 plants spread over a range of crosses that a breeder will have judged from parental performance to be the best.

Breeders will select the best F2 plants on the basis of easily recognized characters such as disease resistance, height and maturity. Further selection on such character is practised upon the F3 generation, after which some breeders will then carry out a preliminary yield trial at the F4 generation and suitability for processing such as baking wheat and malting barley will also be assessed. Yield and processing characters are affected to varying degrees by the environment in which the crops are grown so that the ranking of individuals may vary between environments. This is known as genotype × environment interaction and, to get a more complete picture of the potential of breeding lines, further yields trials are carried out at a number of sites in one or more subsequent generations. At the end of this series of trials, breeders will have assembled a comprehensive body of data upon their remaining selections and will enter the best into official trials. In the UK, approximately 120 winter wheat,winter barley and spring barley lines are entered into National List trials each year in total. National List trials take two years and are conducted according to set protocols in order to select the best three to ten from each crop to enter Recommended List Trials, which are carried out at a number of sites all over the UK. After a year, sufficient data will have been gathered upon the entries to enable a provisional recommendation to be given to between one and five new varieties from each crop each year, although not even one may be widely grown for more than one or two years.

From the above brief outline of a conventional cereal breeding scheme, it can be deduced that such schemes are lengthy, occupying up to ten years to provisional recommendation for the small-grained cereals in the UK and twelve years to make a commercial impact. Bingham and Lupton provide a comprehensive review of a practical winter wheat breeding programme, which is typical of many other cereal breeding programmes. Given the huge investment required to run such a scheme over a 12-year cycle, and the competition provided by others, breeders are keen to exploit alternatives which will either shorten the time-scale or increase the efficiency of identification of elite lines and especially any which combine both elements. The development in molecular biological tools over the past 15 years have led to the possibility of direct selection for the genetic constitution, or genotype, of individuals. This is an attractive alternative as it reduces the amount of selection upon character measurement, or phenotype, when the environment and its interaction with genotype can modify the expression of genes. Conventional plant breeding schemes have been outstandingly successful, contributing an average 1% yearly increase in the yield of barley and wheat. The deployment of molecular biological tools in cereal breeding must either lead to more efficient ways of achieving this rate of progress or lead to increased breeding progress.

Phenotype

The phenotype of a line is the result of the interaction of its genes with the environment. Thus a character with a high degree of genetic control is less affected by varying environmental conditions than one with a low degree of genetic control. Phenotypic selection for performance is reliable in the former case but is likely to be unreliable in the latter. Breeders can therefore make selections for the former group of characters in unreplicated trials or nurseries in the early stages of a programme and reduce their populations to more manageable levels for field trials. Examples of characters that breeders select for in early generations are disease resistance, maturity and short straw. Many of these characters are controlled by a singly gene of large effect (major gene) and are therefore qualitative characters where phenotypic expression is a good indicator of a line's genetic constitution. Most important characters such as yield and quality are, however, controlled by a number of genes, each often of small effect. In such case, because variation is generally continuous and it is impossible to discriminate between the different genotype the characters are termed quantitative. Each individual gene often has a small effect, is subject to considerable environmental modification, and genotype × environment interactions may occur. Accurate prediction of such characters from a single unreplicated trial is extremely unreliable and therefore breeders carry out multi-site trials over two or more seasons to get an estimate of a line's phenotypic potential for character such as yield and quality.

Genotype

The genotype defines the exact genetic constitution of a line but this is unknown for most characters, as the actions of minor genes do not individually produced discernible effects. Indeed, the action of many major genes, such as major gene disease resistance, cannot be recognized without extensive testing with race-specific isolates. This means that breeders are usually completely reliant upon phenotype as a predictor of a line's performance. Knowledge of a line's genotype at individual loci, however, can be used to build up a picture of the relationship between loci. Loci that tend to segregate together are said to be linked and the closer the loci are, the greater the degree of linkage and the fewer the recombinants. Knowledge of such relationships is of value to plant breeders as they can then predict which parental combinations will be most likely to give desirable recombinants. Furthermore, if a breeder wished to find recombinants between particular loci, knowledge of linkage relationship can be used to formulate population sizes necessary to produce as many of such lines as desired. Assembling segregation data on a number of loci enables construction of genetic linkage maps, which can be used in conjunction with phenotypic data to reveal regions of the genome that control a character. These regions are termed quantitative trait loci (QTL) and can be placed on genetic maps

MARKERS

Morphological, Isozyme and Protein

Major gene variation has been known for a long time with many morphological mutants being described. These were used to construct early genetic maps for barley and permitted the addition of major disease resistance genes as they became known. However, many morphological mutants are deleterious and were of little benefit in identifying regions of the genome controlling economically important characters. The effect of some major genes upon other traits could be analysed in populations segregating for the gene. Such studies were used to identify deleterious effects upon yield and quality characters at two dwarfing loci used in spring barley. The deployment of *Hordeum laevigatum* as a source of disease resistance has also led to deleterious associations, probably through introgression of undesirable alleles together with the disease resistance genes. However, such findings became known after the genes had been widely deployed and served to explain observed results rather than advocate the selection of alternative genes. Isozyme markers have also been proposed to augment selection and a number of loci have been included on classical genetic maps of barley, and wheat. Some isozymes have been used to select for specific genes, e.g. the endopeptidase gene on wheat chromosome 7D to select for the VPM1 eyespot resistance gene. Analysis of the high molecular weight glutenin sub-units was found to be a good predictor of bread-making quality of wheat and such analyses are now carried out routinely in wheat breeding programmes prior to baking tests, thus increasing the efficiency of selection. In general, the variation at isozyme loci is insufficient to be of great practical use.

Molecular Markers

Maps constructed from morphological, isozyme and disease resistance genes were the composite picture developed from the integration of a large number of small-scale experiments involving three or four linked loci. The development of methods to assay DNA sequence defferences, of polymorphism, permitted the construction of a map of the whole genome from one single population. This has a number of advantages as a much better estimate of the overall order of loci along the chromosome can be obtained.

Restriction fragment length polymorphism (RFLP)

RFLPs rely on the use of restriction enzymes to digest genomic

DNA where polymorphism can arise through mutations to create or remove restriction sites or by sequence deletions or insertions between sites. After digestion with restriction enzymes, the DNA fragments are separated by electrophoresis and lablelled DNA probes that recognize specific sequences are then hybridized to the fragments and polymorphisms are detected as length differences. Because defferent RFLP alleles are recognized on the same gel, the marker system is said to be co-dominant as heterozygotes (or mixtures) produced both allelic bands.

The initial use of RFLPs in the small-grained cereals was hampered, to some extent, by the relatively low levels of polymorphism that they detected in cultivated germplasm, particularly in wheat. Because RFLP analysis is based upon hybridization assays, comparatively large amounts of DNA of a reasonably high standard of purity are required, which is another limiting factor. A major advantage of RFLPs, however, is the fact that they provide robust anchor loci that are easily distinguished not only across different crosses within a species but also can be used as inter-specific anchor loci. This latter feature has been of great value in comparative mapping within the *Triticeae*, leading to the ordering of homeologous loci across a number of species.

Random amplified polymorphic DNA (RAPD)

The creation of mapping data using RFLPs was laborious, as it required hybridization. The development of molecular marker techniques based on the polymerase chain reaction (PCR) to amplify target DNA segments prior to detection resulted in a quantum leap in the generation of data points. RAPD markers were generated by using random primers in the range of 10-20 nucleotides to detect complementary sites across relatively short distances within the genome. The presence of complementary sites in a genotype resulted in the primer amplifying a DNA fragment under set PCR conditions. A number of different fragments could be amplified with each primer and separated electrophoretically. The absence of a complementary site results in the absence of a DNA fragemts to reveal a polymorphism. This type of marker system is dominant, as a band is either produced or not produced, and the inability to distinguish heterozygotes from a homozygous class is a major disadvantage of the system. Whilst RAPDs could generate data much more quickly than RFLP, they suffered additional disadvantages through difficulties in repeatability, while bands from one cross could not readily be transferred to other crosses. The development of new marker types soon rendered RAPDs obsolete.

Amplified fragment length polymorphism (AFLP)

AFLPs are generated from restriction digests of genomic DNA with enzymes that recognize both frequent and rare sites. This generates a large number of fragments, to which adapters of known sequence are joined (ligated), and then multiplied for several PCR cycles using non-selective primers (pre-amplifications). Selective PCR amplification is then carried out with primers that recognize the adapter sequence plus 1-3 random bases. The amplified fragments are then separated electrophoretically and recognized through the use of radioactive or fluorescent labelling. Like RAPDs, polymorphism is established through the absence of a fragment, which results in an essentially dominant marker system. The more random bases, the fewer the bands. There is therefore a trade-off between the ease of recognizing unique bands on the gel and the amount of data generated. Nevertheless, each primer combination generates a large number of bands, 10-20 of which may, on average, be polymorphic. AFLPs are a much more robust and versatile marker system than RAPDs and their ability of generate a large number of polymorphic bands for each primer combination gives the systems a high multiplex ratio. The use of AFLPs therefore generates a large number of data points in a short time and greatly accelerated the development of genetic maps. *Waugh et al.* compared AFLP loci in three barley crosses and found that they mapped to the same chromosomal locations, so AFLP markers could be used as anchor loci within a species. Precise sizing of AFLP fragments was, however, required to establishment identical loci across several crosses and this was not always possible or practical, limiting the usefulness of the system in comparative mapping.

Simple sequence repeats (SSRs)

SSRs, or microsatellites, are short tandem repeats of mono-, di-, tri- and tetranucleotides although more complex repeats have been detected. They are abundant in mammals and have been found in plants by searching sequence databases. They can also be found by sequencing random clones of genomic DNA. The main advantage of SSRs over RAPDs and AFLPs is that they are, like RFLPs, co-dominant markers and can therefore reliable identify heterozygotes. They have also proved to be multi-alleleic, with over 30 different alleles being detected by a single SSR primer pair in barley. This facet renders then highly informative so they have high polymorphic information content (PIC) values. The main drawback of SSRs is their high development cost, requiring extensive DNA cloning and sequencing for their identification. The production of libraries enriched for specific SSR motifs has been found to greatly increase the rate of SSR discovery but the process is still resource hungry. A number of groups have invested resource in developing SSRs and 230 are available for wheat and over 560 for barley. Due to their high PIC values, SSRs are proving to be excellent anchor loci for comparing various studies and are the current marker system of choice within a species. Primers for SSR loci developed for one species, however, usually fail to amplify a product in another species and, even if a product is produced, it cannot be relied upon as a homoeologous locus.

Derivative marker systems

A number of marker systems have been developed that share features of the other PCR based systems but differ in a number of respects. Two methods based on simple sequence repeats are inter-simple sequence repeats (ISSR) and *Copia*-SSR, also termed retrotransposon-microsatellite amplified polymorphism (REMAP). Two other methods, which are based on retrotransposon, are sequence-specific amplified polymorphism (S-SAP) and inter-retrotransposon amplified polymorphism (IRAP). All four are PCR based and produce a number of polymorphic bands for each PCR reaction. Results published so far suggest that all have a higher multiplex ratio than AFLPs, e.g. ISSRs in rice, S-SAPs in barley, IRAP and REMAP in barley. All are dominant marker systems and are therefore less informative than SSRs and their use as an anchor loci has not been evaluated, but they do seen to have same promise as an alternative to AFLPs.

Single nucleotide polymorphisms (SNPs)

SNPs occur as a result of a single base substitution, deletion or adition in a sequence and have been detected at a rate as 1 in 26 base pairs in barley. The advantage of SNPs is that specific oligonucleotides can be designed to detect a particular polymorphism in a positive or negative fashion. One can then analyse a batch of material with a high throughput and automated assay. Such a system also offers the possibility of dispensing with gel-based detection methods, which are one of the current limiting factors in marker analysis. As more sequence data become available, particularly from the commitment of the public and private research communities to develop large libaries of clones representing the expression of single genes (expressed sequence tags - ESTs), it is becoming more realistic to identify SNPs on a large scale in the small-grained cereals. At present, there is no simple, robust way of detecting SNPs other than amplifying a target sequence for a test panel of genotypes and comparing all the sequences to reveal polymorphisms. This therefore requires a heavy investment in DNA sequencing but is now feasible with the latest generation of sequencers. Although SNPs have been found in barley which are diagnostic for the *rym4* and *rym5* resistance genes to Barley Yellow Mosaic Virus, it is likely that value in diagnostics. The genotypic constitution at a number of successive loci (haplotype) gives much more information than the genetic constitution at an individual locus as the possible combinations are much more than the two at a single SNP locus. It is therefore likely that SNPs will become the main marker system in diagnostics but their value in generating genetic maps is unknown as yet, although their development would mean that maps would be based on known functions rather that largely anonymous markers.

Genetic Maps

Genetic maps show the location of genes on chromosome and the distances between them so that one can estimate whether or not two genes are likely to segregate together or, if breakage of a linkage is desired, the population size necessary to ensure recombinants between two neighbouring loci. Genetic maps are constructed by assembling segregation data for a series of markers, which can be used to estimate the location of genes relative to each other. With morphological markers, segregation could usually only be studied for a few loci at a time. The standard procedure is to compare the frequencies of the various parental and recombinant classes in a segregating population such as an F2 or, more recently, doubled haploids. If two loci are completely unrelated, then the observed frequencies should be within the sampling error of that expected for unlinked loci. If the observed frequency of the recombinant classes is significantly less than that expected, then the loci are located in the same region and the degree of linkage can be estimated from the observed frequencies of the various marker classes. Thus, placing a new marker gene on a gentic map was a matter of crossing it to a series of other known marker loci and checking the segregation of 38 loci across seven linkage groups in barley by 1951 and nearly 80 loci were assigned to the seven barley chromosomes by 1962. Methodologies were developed to combine segregation data from a series of separate experiments to produce more inclusive maps which, for barley, were published annually in the *Barley Genetics Newsletter*

Mapping in this way was very much a matter of trial and error and it was not always possible to locate commercially important major genes. For examples, the *Hordeum laevigatum* mildew resistance gene (*MlLa*) was found in the spring barley cultivars Vada and Minerva and many of their derivatives but a comprehensive conventional linkage analysis failed to localize the gene to any one barley chromosome. The development of molecular maps soon enabled the gene to be located on chromosome 2H.

Associations of major genes

The assignment of some commercially important major genes to particular chromosomes helped to elucidate some information about the control of quantitative characters. By developing random inbred populations that segreated for an easily characterized major gene, it is possible to assess the whole population for a range of economically important characters. The variation for each character can then be partitioned into that acribable to differences between the alternative alleles at the major gene locus and the variation within groups. If the former is significant when tested against the latter, then the region of the genome in which the major gene is located also plays a role in the genetic control character measured. This method was used to fine that the *sdwl* and *ari-eGp* dwarfing genes in barley were associated with a number of other characters. Whilst this analysis gave an indication of the possible consequences of deplying specific major genes, it is essentially retrospective and cannot distinguish whether associations are due to linkage or pleiotropy.

Mapping software

The development of large arrays of molecular marker data meant that manual mapping methodology was far too labour intensive. Computer software was therefore developed to reduce the amount of manual calculation. The principles behind mapping remain the same but were updated to examine a large number of marker combinations simultaneously. The most popular (by citation) mapping software is MAPMAKER, although the more recent JOINMAP has gained in popularity, again by citation. GMENDEL has also been used and all three differ in their approach to map construction but essentially produce similar maps from a given set of data. All have a reasonable user interface, are well documented and each has routines to evaluate the quality of the input data and/or the maps produced. They can all cope with various different types of populations generally used in mapping (F2, backcross (BC), recombinant inbred lines (RIL) and doubled haploid (DH)) inbreeding species such as the small-grained cereals. JOINMAP has the added advantage of being able to develop a composite map from different mapping populations but with some common reference markers.

Physical maps

Genetic maps provide an interpretation of the distribution of genes along chromosomes based purely upon recombination. Using suitable stains, banding patterns can be directed observed on barley chromosomes. Some of these bands differ between cultivars and can be scored as genetic markers in segregating populations. Linde-Laursen combined a banding pattern marker with other markers to demonstrate that the genetic map location of the band differed from its physical location. This reflects the tendency for recombination to occur at the ends of cereal chromosomes, rather than at random. The development of fluorescent in *situ* hybridization (FISH) provides an opportunity to vusualise loci directly upon chromosomes and therefore extend the comparison of physical and genetic maps. Laurie *et al.* used ribosomal RNA markers to demonstrate considerable differences between the physical and genetic maps of the long arm of barley chromosome 2H. Kunzel and Korzum found no physical markers in some 70cM of the distal part of the long arm of barley chromosome 3H, indicating that the genetic map was considerable lengthened in comparison to the physical. Differences between genetic and physical maps have important implications for those studying the organization of the genome but the genetic map represents what it is possible to achieve through recombination in terms of segregation of markers and it therefore the most relevant measure for those studying the inter-relationships of characters.

Published maps

In the early 1990s extensive maps were published for two barley populations, Proctor × Nudinka, and Igri × Franka. Later the North American Barley Genome Maping project developed extensive maps for two barley crosses, Steptoe × **Morex**, and Harrington × TR306. The three ancestral genomes of bread wheat necessitated the construction of three sets of maps, with many homoeologous loci that map to two or three of the constitutent genomes. The early use of RFLPs in wheat mapping was to construct maps from different crosses for each linkage group (e.g. Chao *et al.*).

The first genome-wide maps of wheat crosses involved its wild relatives but, because of the problems of low levels of polymorphism detected by RFLPs an, RFLP-based map of an inter-varietal bread wheat cross took longer to develop. Similarly, the first oat RFLP map to be published was for a cross between wild species with a map from cultivated oats being developed later. In addition, JOINMAP has been used to construct composite maps of barley over four and seven mapped populations. RAPDs were mainly used to fill the gaps in maps, using known RFLP loci as points to anchor the data to chromosomes. Several barley maps including RAPD markers were published, e.g Vogelsanger Gold × Alf and Blenheim × E224/3. Becker *et al.* first reported the use of AFLPs in barley mapping and Powell *et al.* and Qi *et al.* reported other major barley mapping efforts using AFLPs. A largely AFLP-based map of a bread wheat cross has also been developed and AFLPs have also been used in oat mapping. SSRs are now being incorporated into genetic maps of barley and maps of bread wheat based on SSRs have also been published. The most efficient current mapping strategy is to combine a number of reference SSRs and AFLPs and/or the derivative markers noted above to fill in the gaps.

All of the above maps are notable for their relative lack of non-molecular markers. In the barley Proctor × Nudinka map, a major gene mildew resistance at the *Mla* locus on chromosome 1H and the naked locus *n* on 7H is presented. The lack of non-molecular markers reflects the lack of major gene morphological variants in crosses between cultivars, be they of barley, oats or wheat. More recently, attempts to incorporate morphological markers on barley molecular maps have been made through the Oregon Wolfe Barleys (http:// www.ccs.ort.edu/barley/wolfebar/wolfnew.htm) and Kleninhofs *et al.* have attempted to place some barley morphological markers on molecular maps.

With the large and ever-increasing numbers of molecular

markers available for mapping in barley, oats and wheat, markers are available that give comprehensive genome coverage of all species. The problem is now establishing an order amongst a huge range of data. It is important to remember that the statistical resolution of genetic distances is dependent upon the number of individuals in the population being measured, as can be seen for the linkage formulae given by Mather. Whilst mapping software usually estimates orders of loci to tenths of a centi-Morgan, these are beyond the resolution of most of the populations being studied and thus the fine-map order of a population of say 100 backcross or doubled haploid individuals cannot be relied upon. To solve this presentational problem, Kleinhofs *et al.* have developed a system of 'bins' at stratified intervals along the chromosome. This has the advantage of obviating the need to integrate the present maps with huge numbers of data-points on them.

CHARACTERS

One of the biggest benefits of having genetics maps is being able to localise genetic control of quantitative traits to specific regions of the gonome, which requires genotyping and phenotyping of a random population of lines from a cross. For the same reasons that it is difficult for a plant breeder to select for important quantitative characters such as yield and quality, it is desirable to carry out phenotypic evaluation over a number of different environment. Thus, the phenotyping component is not trivial and ideally requires a population of inbred lines, either recombinant inbred lines (RILs) from a selfing series or doubled haploid (DH). Prior to the advent of molecular markers, manipulation of pairing in wheat genomes had been carried out to develop special chromosome stocks that carried substitutions of whole chromosomes of chromosome segments. Phenotypic analysis of these stocks enabled quantitative and qualitative, variation to be firstly located to a chromosome and then to more specific regions. This elegant series of experiments, reviewed by Law *et al.* allowed much progress to be made in the genetic analysis of wheat. The development of the stocks was laborious and limited the range of germplasm that could be surveyed. The development of molecular maps meant that quantitative traits could be studied not only in a wider range of germplasm but also in more species.

QTL Detection

Having assembled a set of genotypic and phenotypic data for a cross, the simplest way of detecting QTL is,. for each marker in the genotype test, to classify the phenotypic data for each character into the parental marker groups. The means of each marker group can then be compared and tested for significance by analysis of variance or regression. Edwards *et al.* detected this approach is that it is laborious and only gives an association of marker loci with a character. In cases where neighbouring marker loci are far apart, the differences between the marker groups may not reach significance for a QTL of small effect and/or the QTL may be some distance from the marker.

Interval mapping was developed to step along a genetic map at set intervals and, at each interval, test for the presence of a QTL. The test is based upon the phenotypic means of the marker classes and the distance between the markers. This approach gives a more precise location of QTLs and MAPMAKER/QTL software has been developed to automate the procedure. The approach used maximum likelihood and could efficiently detect singly QTL effects on a linkage group. In the presence of more than one QTL per linkage groups, the method could either fail to detect any effect at all, if the loci from a parent were of opposite sign, or detect a 'ghost' QTL. Haley and Knott presented a least squares approach to interval mapping that gave similar results to maximum likelihood but was simpler to apply, could be adapted to a range of situations and was robust enough to detect multiple QTL in a linkage group.

QTLs of large effect mask others of small effect and Jansen proposed the use of co-factors to account for variation in other regions of the genome when scanning a target region. Jansen ans Stam proposed using the whole marker set as initial co-factors and then eliminating them in a backward elimination regression procedure but this can lead to over-parameterisation. Hackett proposed a forward selection method to identify co-factors and this has been adpoted by Utz and Melchinger. Similarly Zeng proposed the use of marker co-factor to account for the effect of one QTL when searching for another. The use of marker co-factors has been termed compound interval mapping and has been found, both by simulation and application to real data, to be more efficient at detecting QTL and also increases the precision of location. Software packages which implement this approach are MapQTL. QTL Cartographer and PLABQTL MQTL uses backgound markers to account for QTLs in regions other than that currently being scanned in an approach termed 'simplified compound interval mapping'. PLABQTL and MQTL can also analyse data for a trait from more than one environment and determine which QTLs and which are effective in one or more environment (QTL × environment interactions). PLABQTL differs from MQTL in its approach to this problem as it looks for QTL in the overall means from different experiments and then attempts to fit this model to each environment in turn. This means that, as MQTL searches the accumulated data over all environment for QTL, it is more likely to detect QTL × environment interactions of the cross-over type in addition to interactions of magnitude. Compound interval mapping is now the method of choice for QTL detection and choice of programme depends on objectives, availability and ease of use and interpretation.

Marker regression has also been proposed as a method of QTL detection. This approach uses all markers in a linkage group to detect the presence of possible QTL and deviations of the observed values from predicted could indicate the presence of more than one QTL in a linkage group. This is a simple and easy to implement approach which has value in some situations but it does not detect QTL × environment interactions.

QTLs Reported

A number of QTL studies in barley, oats and wheat have been published, largely centred heading, height, yield and disease resistance. Some studies looking at malting quality parameters in barley and processing quality characters in wheat have also being published. Few comprehensive studies of agriculturally important traits have published as most concentrate upon one or a few traits at a time. Barley appears to be an exception in that crosses have been studied for a range of agronomic, disease, yield and quality characters in North American, European and Australian germplasm. No comparable studies have been published in either oats or wheat as most of the results in the public domain have concentrated on one or a few characters. For example, separate studies in wheat have focussed on QTLs for plant height, disease resistance, flour viscosity and frost tolerance. There are fewer still published reports for oats but QTLs for resistance to Barley Yellow Dwarf Virus and groat oil content have been detected.

In a number of the QTL studies conducted in barley, a major gene with a marked agronomic effects has been segregating, either a dwarfing or a disease resistance gene. In many cases, the major genes have been associated with QTLs for other characters. For example, the *sdw 1* dwarfing gene on chromosome 3H in spring barley has been associated with malting quality characteristics. There are also a number of instances where QTL 'hot-spots' have been detected. For instance, the distal region of the short arm of chromosome 5H in Harrington × TR306 is associated with QTL for yield, maturity and some malting parameters amongst other characters. The confidence intervals of QTL are so large that it is still not possible to determine whether association with major genes or 'hot-spots' are due to genuine pleiotropy or to tight linkage. The construction of special genetic stocks in which a series of isolines differ by small, neighbouring fragments of the genome should help to resolve this question.

QTL Validation

Before utilising QTL information in a breeding programme, some form of verification of effect is desirable. This problem has been addressed in several practical studies in barleys. Hayes *et al.* identified two QTL which, when combined, accounted for over 25% of the phenotypic variation in malt extract and some other malting quality parameters in the Steptoe × Morex mapping population. Han *et al.* found that, in another sample of lines from

Steptoe × Morex, the two QTL accounted for much less of the phenotypic variation for the characters. Romagosa *et al.* came to the same conclusion studies of QTL effects in maize and concluded that accurate estimation of QTL effect requires extremely large populations. Working with populations of 100-200 lines resulted in considerable bias in estimate of effect. Bias in estimation of QTL effect appears to be a genuine problem but need not detract from results already accumulated, provided that the methodology identifies the most important QTL.

DEPLOYMENT OF MOLECULAR MARKERS

Varietal Identification

Ainsworth and Sharp proposed that RFLPs could be used in varietal identification but, as noted above, the overall level of polymorphism detected by RFLPs is relatively low and the technique is demanding. Terzi demonstrated that RAPDs could also be distinguish between varieties of barley, oats and wheat but the disadvantage of this method is that the portions of the genome being sampled are unknown. Due to their high multiplex ratio, AFLPs are another possibility for cereal varietal discriminaiton. The dominant nature and low polymorphic information content of AFLPs may, however, make them less useful in distinguishing between closely related cultivars and, like RAPDs the genomic distribution of the markers is likely to be unknown. SSRs appear to be particularly useful for varietal discrimination, due to their multi-alleleic nature and high PIC values, and can provide unique genetic finger-points of crop cultivars. Russell *et al.* used a panel of 11 SSRs to distinguish between 24 winter and spring barleys either on the UK National or Recommended Lists. In fact three different subsets of four SSRs were shown to be capable of differentiating between all 24 barleys. It should be noted that two of the winter barleys were derived by the same breeder from the same cross and were morphologically similar but differed at 4 of the 11 SSR loci surveyed.

Germplasm Analysis

Cultivated germplasm

Molecular markers have a number of attractions for determining the genetic variability in a germplasm pool and, with the development of genetic maps, the level of polymorphism detected can be related to genomic regions. A multi-variate technique called 'principal co-ordinate analysis' can be applied to data collected from a panel of genotypes to reveal the pattern of relationships between them. This relationship can be inspected graphically either by plotting the scores for the first two or three principal coordinates or, if they fail to account for a sufficiently large portion of the variant, a dendrogram. For barley, RFLPs have been used to differentiate between elite winter and spring European cultivars. Within these groups, discrimination between either two- and six-rowed types is possible amongst the winter cultivars and between feed and malting cultivars amongst the spring cultivars.

RAPDs and AFLPs have also been used to examine the relationship between barley cultivars with similar results to those obtained from RFLPs. As for varietal discrimination, the disadvantages of using RAPDs and AFLPs is not knowing the chromosomal location of the markers. SSRs have a number of advantages in studying genetic relationships between genotypes as the multiple alleles that can be detected at any one locus give a better overall impression of the variation. Amongst European spring barley, Russell *et al.* have shown that over 70% of all the alleles detected by a panel of 28 SSR loci distributed across all seven barley chromosomes were found in 17 'founder' genotypes which were largely old land-races, their early derivatives or disease resistance donors. Furthermore, more modern genotypes, as encompassed by a group released since 1985, possessed only 35% of the total alleles. This confirms the narrowing for the genetic base of modern barley, which Fischbeck suggested was due to:

- the exploitation of relatively few land-races in plant breeding from the 1880s to 1920s.
- a small number of cultivars that represented a significant advance and consequently featured heavily in subsequent breeding programmes
- the use of exotic germplasm largely as a source of disease resistance.

Whilst this dependence on 'founder' genotypes is fairly universal across the genome, Russel *et al.* found that post-1985 spring barley cultivars did contain some novel alleles, notably in a region on chromosome 5H that may reflect selection for yield and/or quality.

Molecular analysis of wheat germplasm has not progressed as far as that of barley. The relatively low level of polymorphism detected by RFLPs in wheat hindered progress and RAPDs were found to be unreliable due to primers amplifying different non-homologous sequences between genotypes. Paull *et al.* used a panel of mapped RFLPs to analyse a collection of Australian bread wheat genotypes to produce four major groupings that could be associated with the origin of the lines. They also found a number of differences between the observed and expected genotype based upon pedigree information. They concluded that, provided that the markers gave a broad, evenly-spaced genome coverage, the use of markers to estimate similarity was more meaningful than pedigree information as the latter did not account for selection and drift. Paull *et al.* used the low level of polymorphism detected by RFPLs to argue that, where differences existed, they might well reflect the effects of phenotypic selection. In contrast, multivariate analysis of a sample of 11 winter wheat cultivars from Austria and Germany based on pedigree information, RFLPs, AFLPs and SSRs did not produce any meaningful groupings, nor were there any common trends between the four dendrograms. Over all the pairwise combinations of the 11 genotypes, there were no significant correlations between the pairwise combinations of genetic similarities estimated by the three types of molecular markers and only the estimate produced by AFLP showed a significant correlation with pedigree.

Working with oats, O'Donoughue *et al.* used principal co-ordinates analysis of RFLP data scored on a panel of 84 cultivars. They found two major groupings from the analysis, which generally corresponded to winter and spring cultivars, and that some sub-groups could be associated with breeding history. Whilst they found a significant correlation between genetic distance, as measured by RFLPs and pedigree, like the studies of wheat and barley noted above, the correlation was small.

Exotic germplasm

ISSRs have been used not only to discriminate between winter and spring cultivated European barleys but also to highlight the distinctiveness of wild barley *Hordeum spontaneum*. This has been explored further with the used of mapped SSRs, which has highlighted genetic bottlenecks that have occurred at specific regions of the barley genome, notable on chromosome 7H. A survey of land-race material collected by ICARDA showns that it is intermediate between *Hordeum spontaneum* and cultivated barley in terms of allelic diversity. Greater diversity exists in the land-races than in *Hordeum spontaneum* in specific regions on the genome, much of it from novel alleles.

In *Hordeum spontaneum,* Pakniyat *et al.* analysed singly accessions from 39 collections sites with AFLPs derived from 12 primer combinations. They found that the first three principal co-ordinates grouped the accessions according to the eco-geographic variable associated with the collection sites. The first principal co-ordinated differentiated between the sites on the basis of altitude whilst the second and third discriminated between longitude and latitude respectively. If one could then identify the phenotype that is favoured at a particular site, then such an analysis provides a powerful method of studying the direct action of natural selection.

Pedigree Analysis

The multi-allelic nature of SSRs makes them extremely useful in analysing prdigrees of genotypes. Swanston *et al.* found a barley SSR, Bmac213, was linked within 7 cM of a major gene locus affecting non-production of epiheterodendrin (a cyanogenic glycoside precursor important is some distilleries). They were able to trace the origin of this alleles back through the pedigree of the spring barley cultivars Derkado the cultivar Emir. Other derivatives of Emir were also non-producers of epi-heterodendrin and also possessed the same Bmac213 allele as Derkado. The origin of the Emir allele was probably the disease resistance donor

Arabische, although the accession tested neither possessed the Derkade Bmac213 allele, nor was it a non-producer of epi-heterodendrin. This was probably because the accession used in the breeding of Emir differed from that in the collection studied. This illustrates the general problem that material in collections, particularly less homogeneous material, may not accurately represent the variation in a particular genotype due to genetic drift and/or difficulties in ensuring that all the variation is passed on from generation to generation during multiplication and maintenance of accessions.

Marker-assisted Selection

With the development of molecular maps, there are now many published associations of characters with markers that are close enough to be used in marker-assisted selection programmes (for a review of those published in wheat see Gupta *et al.*). Theoretical studies have highlighted the benefits of applying marker-assisted selection in breeding programmes of the small-grained cereals which suggest that marker-assisted selection should offer a substantial advantage over phenotypic selection for characters of low heritability under intense selection. The problem then becomes one of identifying robust QTLs that can be used in marker-assisted selection for such characters. Various selection strategies, including marker-assisted selection, have been compared for malt extract and some other malting quality characters and yield in barley. Both studies demonstrated that marker-assisted selection was effective, particularly when combined with phenotypic selection in the later stages of a breeding programme. A strategy of using genotypic selection to assemble a pool of 'elite' germplasm in which a number of desirable QTLs of major effect for traits that are difficult to measure followed by phenotypic selection of identify the best of these lines has a number of advantages:

- Selection is concentrated on lines that are most likely to meet the desired standard for yield and quality.
- The number of lines in the advanced stages of breeding programmes is reduced.

- Response from genotypic selection of QTL of small effect is, at best, likely to be negligible and therefore not cost-effective.

Marker-assisted selection is commercial breeding of wheat and barley is carried out, but generally for qualitative characters that are difficult to assess phenotypically. Markers can also be used to gain some knowledge about unknown germplams, for example to choose parents for use in crossing. Alternatively, where progeny are known to be segregating for a qualitative character, marker-assisted selection may be applied at varying stages of the selection programme to identify those carrying the target gene. The Barley yellow Mosai Virus complex is one example where molecular markers have been deployed. The *rym4* resistance gene on barley chromosome 3HL was found to be flanked by the RFLPs MWG10 and MWG838. The latter was converted into a sequence tagged site (STS) and Tuvesson *et al.* demonstrated its use in molecular-assisted selection. The SSR Bmac29 is also linked to the *rym4* locus and is capable of differentiating not only between resistant and susceptible alleles not also between the *rym4* and *rym5* alleles.

Backcross Conversions

The benefits of using marker-assisted selection in backcrossing have been demonstrated theoretically. Toojinda *et al.* identified a source of barley stripe rust (*Puccinia striiformis*) resistance in exotic germplasm and then used a backcross conversion scheme to introgress the resistance into the locally adapted cultivar Steptoe. Progeny from the first backcross that carried the desired genotype were selected by using RFLP markers that were polymorphic and flanked the QTL. The selected BC1F1 genotypes will be heterozygous for the target segment and therefore doubled haploids were produced from them to produce lines. The doubled haploids were then re-selected with the flanking RFLP markers to identify those that carried the QTL. AFLPs were used to screen the backgound genotype and showed that the percentage of the donor genome varied from 7 to 60%. Under a normal backcrossing scheme, the percentage of the donor genome is expected to average 25% in a first backcross so some of the lines approached the equivalent of a third backcross. The capacity to identify such lines demonstrates the power of marker-assisted selection in a targeted backcrossing strategy. Selected backcross inbred lines can be produced within a two-year time-scale, regardless of whether the crop is winter- or spring- sown. The scheme is most suitable for one or two loci and , because it relies on selection of heterozygotes at the marker loci, works best with co-dominant markers for the target locus. If working with exotic germplasm, the relatively low levels of polymorphism detected by RFLPs should not be a problem. Given the increasing numbers of loci that cover the genomes of barley and wheat and the other advantages noted above, SSRs are the current marker of choice in targeted backcrossing schemes.

FUTURE PROSPECTS

Candidate Genes

Expressed sequence tags (ESTs) are developed by partially sequencing cDNA clones from target tissues. From the sequence information collected, searches can be made against publicly available databases and, given an adequate level of sequence homology, some idea of function can be derived. A big international effort is being made to assemble comprehensive EST databases for the *Triticeae* that will be publicly available. At the same time, private companies are investing heavily in developing EST libraries for their 'core' crops but it is unlikely that much of their information will be made public. The challenge in both the private and public sector is to find ways of incorporating ESTs on genetic maps in order to determine whether any correspond to QTSl of interest, which would facilitate the change in emphasis in genetic maps from largely anonymous to known-function markers. Combining this information with QTL mapping will not only enable some functionality to be assigned to a particular QTL but also provide sequence data for its direct manipulation. Current QTL mapping methodology is, however, far too imprecise to enable such assignations to be made unambiguously. Mappings large EST libraries also posses a number of problems, including choice of the most appropriate marker system. Some 5% of barley

EST sequence have been found to possess SSRs but their informativeness has yet to be established. The identificaiton of SNPs in ESTs appears to be more informative but identification currently requires a massive sequencing effort to reveal differences in panels of genotypes.

DNA Chips

DNA sequences can now be assembled in micro-arrays over a small area -so- called DNA chips. DNA chips are being advocated as suitable for a number of applications, including genotyping where they would be highly suitable for the detection of bi-allelic markers such as SNPs. The high level of automation that is possible with DNA chips means that it would be possible to carry out high through-put genotyping, which would enable marker-assisted selection of a large scale. DNA chip technology is still under development and its cost is not yet apparent. Whilst it appears promising, it will have to show a clear benefit over current approaches before it becomes a worthwhile investment purely for germplasm screening purposes. Apart from some obvious qualitative characters, the problem of identifying QTLs that are robust enough to warrant application of marker-assisted selection (noted above) will limit the amount of screening that could currently be done using DNA chips. If the technology was sufficiently cost-effective, then it may, however, be suitable for screening QTLs that are less robust.

Bioinformatics

The development of SSRs for a wide range of crop species has led to the possibility of not only examining genetical relationships through principal coordinates analysis but also using the map locations of the markers to produce a more meaningful graphical representation of the allelic constitution of genotypes. Computer software has been developed to construct these graphical representations, e.g. Supergene, but no package published so far can cope with the range of alleles that can be generated at one locus by SSRs. As SSRs are integrated into mapping studies of qualitative and quantitative characters, it becomes possible to assign values of specific SSR alleles. A variety of crossing strategies could then become possible by genotyping parental lines. Working within adapted germplasm, one could:

- assemble all the favourable QTL alleles for important characters that are difficult to screen for in cross combinations or
- make crosses between parents that both possess the major favourable QTL alleles and use conventional phenotypic selection to assemble the minor QTL alleles in a favourable combinations.

Combining the information in such a database with genotypic information from the unadapted gene-pool would facilitate the identification of novel alleles in the region of important QTLs. These novel alleles may also represent novel variation for the character and can be introgressed through a marker-assisted backcrossing scheme to test whether or not they represent useful variation for deployment in plant breeding. Useful and novel phenotypic variation in wild relatives has already been detected in tomato and rice, although it remains to be throughly tested whether the QTLs from wild relatives represent new loci or variants at existing loci. The approach presented by Tanksley *et al.* is interesting as it combines mapping with introgression in a methodology termed 'advanced backcrossing'.

Recombinant Chromosome Substitution Lines (RCSKs)

One of the problems in establishing reliable QTL locations is that the estimate is usually based on very few recombinant lines and therefore it is not surprising that QTL effects vary in different studies, even when sampling from the same cross. One solution to this problem is to create a series of recombinant chromosome substitution lines. These are essentially a series of isogenic lines which differ only for a small (ideally contiguous) segment of chromosome from a donor line. These can then be replicated and widely tested along with the recipient cultivar to estimate the effect of individual segments in isolation. This approach can be applied in a sequential manner by an initial screen of lines with relatively large introgressed segments to identify candidates for further rounds of backcrossing to break up a segment into a series of smaller ones. This approach can therefore be used to derive more accurate estimates of QTL effect and possible to refine the target segment to be used in marker-assisted selection. Another benefit of this approach is that it should help to distinguished between close linkage and pleiotropy, as well as providing material for gene cloning.

CONCLUSIONS

Molecular markers have undoubtedly revolutionized the genetic analysis of performance traits and can be applied in marker-assisted selection schemes. Markers have been successfully applied in plant breeding but there are few published examples. Those that have been published are largely concerned with the manipulation of qualitative characters that give a characteristic phenotype. The identification of QTL influencing important quantitative characters does not yet seem reliable enough to warrant the use of markers tagging QTLs in a marker-assisted selection programme. What is certain is that technological developments will continue and much more information about the genetics, biochemistry and physiology of crop plants will emerge. The challenge is, and will continue to be, to integrate all the information with conventional phenotypic selection into efficient plant breeding programmes.

5

Assesment and Legislative Issues

GM CROP DEVELOPMENT

Since the first field trials of transgenic crops were conducted in the USA and France in 1986, there has been a rapid growth in activity with field trials being carried out globally involving at least 56 different crop species. In 1999 the acreage of GM crop plants grown for commercial purposes world-wide was expected to reach 73 million acres, with crops grown mainly in the USA and Canada.

TABLE 5.1 : RELEASES OF GENTICALLY MODIFIED ORGANISMS PER COUNTRY 1998

Country	%	Country	%
USA	70.45	Sweden	0.37
Canada	11.83	New Zealand	0.34
France	4.72	Denmark	0.31
Belgium	2.02	Brazil	0.28
UK	1.84	South Africa	0.17
Italy	1.71	Finland	0.11
Holland	1.47	Portugal	0.06

Spain	1.20	Russia	0.06
Japan	1.17	Bulgaria	0.05
Germany	0.89	Austria	0.03
Australia	0.88	Switzerland	0.03

While Europe has led the way in terms of GM crop development and evaluation, the commercial situation, in the UK in particular, is very different. To date the UK has approved 135 applications for release, but for reserch purposes only. Whilst there are an incerasing number undergoing experimental and performance trials, no contents for release for commercial purposes have yet been granted. Commercialisation of the first GM variety is under review in response to mounting public opposition and demands for a five-year freeze until further experimental analysis satisfies concerns about GM crop safety.

TABLE 5.2 GENETICALLY MODIFIED PLANT SPECIES

African violet (*Saintpaulia ionantha*)
Alfalfa (*Medicago sativa*)
American Chestnut (*Castonea dentata*)
Apple (*Malus domestica*)
Asparagus (*Asparagus officianalus*)
Barley (*Horedeum vulgare*)

Beet (*Beta vulgaris*)
Belladonna (*Astropa belladonna*)
Broccoli, cauliflower and cabbage (*Brassica oleracea*)
Forage rape (*B.oleracea* var. *acephala*)
Kale rape (*B. oleracea* var. *biennis*)
Brown mustard (*Brassica nigra*)
Carnation (*Dianthus carophyllatus*)
Carrot (*Daucus carotta*)
European Chestnut (*Castanea sativa*)
Chicory (*Cichorium intybus*)

Chrysanthemum (*Chrysanthemum morifolium*)
Cotton (*Gossypium hirsutum*) (7%)
Maize (*Zea mays*) (38%)
Marigold (*Tagetes sp.*)
Melon (*Cucumis melo*)
Mustard (*Brassica juncea*)
Oat (*Avena sativa*)
Oilseed rape (*Brassica napus*) (13%)
Onion (*Allium cepa*)
Orange (*Citrus sp.*)
Papaya (*Carica papaya*)
Pea (*Pisum sativum*)
Peanut (*Arachis hypogaea*)
Pelargonium sp.
Pepper (*Capsicum annum*)
Pine (*Pinus sp.*)
Pineapple (*Ananas comosus*)
Poplar (*Populus sp.*)
Potato (*Solanum tuberosum*) (12%)
Rice (*Oryza sativa*)
Rose (*Rosa hybrida*)
Silver Birch (*Betula pendula*)

Cranberry, European (*Vaccinium oxycoccus*)
Creeping bentgrass (*Agrostis stolonifera*)
Cucumber (*Cucumis sativus*)
Cucurbita texana
Cucurbita pepo
Currant (*Rubus idaeus*)
Eggplant (*Solanum melanogea*)
Ethiopian mustard (*Brassica carinata*)
Eucapyptus (*Eucalyptus camaldulensis*)
Flax (*Linum usitatissium*)
Gladiolus sp.
Grape (*Vitis vinifera*)
Kentucky Bluegrass (*Poa patensis*)
Kiwi fruit (*Actinidia deliciosa var. deliciosa*)
Lettuce (*Lactua sativa*)
Lisianthus (*Eustoma grandiflorum*)
Lupin (*Lupinus angustifolius*)
Spruce *Picea sp.*
Spruce, Norway *Picea abies*
Sorghum (*Sorghum bicolor*)
Sugar beet (*Beta vulgaris*) (2%)
Sugarcane (*Saccharum officinarum*)
Sunflower (*Helianthus annuum*)
Sweet potato (*Ipomoea batatas*)
Sweetgum (*Liguidambar. sp.*)
Tamarillo (*Cyphomandra thaliana*)
Thale cress (*Arabidposis betacea*)
Tobacco (*Nicotiana benthamiana*)
Tobacco (*Nicotiana tabacum*) (5%)
Tomato (*Lycopersicon esculentum*) (10%)
Turnip rape (*Brassica rapa*)
Walnut (*Juglans sp.*)
Watermelon (*Citrullus lanatus*)
Wheat (*Triticum aestivum*)
White mustard (*Sinapsis alba*)

Those species comprising the majority of releases are indicated by the relevatn percentage of releases within the OECD.

The state of GM crop development in the UK can be summarized as follows: oilseed rape and maize are nearest to commercialisation; modifications include varieties tolerant to the herbicides glucosinate ammonium (Challenge variety) and glyphosate (Roundup variety). In addition oilseed rape varieties modified for expression of improved oil quality such as those expressing a high lauric acid content are also close to the market place.

A wide range of GM crops are currently in experimental trial including spring wheat (disease resistance), sugar beet (herbicide tolerant and altered carbohydrate metabolism), potato (altered carbohydrate, virus resistance) and maize (herbicide tolerant and insect resistance). Genetic engineering has also enabled higher yielding hybrid systems to be produced by the development of GM male sterile plants, a number of which are currently being tested for yield and overall performance. Cultivars of spring and winter oilseed rape, sugar beet, fodder beet and forage maize are currently being assessed in the UK's statutory National List (NL) trials. Inclusion of a variety onto the NL and the EC common catalogue is an essential precursor to commercialisation.

Crop development in the future is likely to continue with the production of varieties with improved pest and disease resistance. These developments are also likely to include plants compatible with effective weed control and environmentally friendly farming methods, and crops with tolerance to salinity, drought or frost. There is also likely to be an increasing emphasis on the development of varieties that are bred for processing purposes such as the production of novel oils, starches and high-value pharmaceutical compounds, for example, vaccines. GM crops are likely to be an important future development in providing a substitute for fossil fuels. There is good potential for utilizing genetic engineering for the synthesis of plant-based alternatives to fossil fuels using a range of widely grown oilseed crops like oilseed rape to produce economically viable quantities of oil. GM crops may also provide substitutes for other non-renewable resources from which we derive, for example, many industrial oils used in the manufacture of plastics, detergents, inks and lubricants. One example is the isolation of the genes encoding the enzymes responsible for the synthesis of petroselinic acid, a fatty acid with potential for use in making detergen and nylon polymers. Further emphasis will also be placed on the development of designer health crops, resulting in better tasting, nutritionally enhanced or 'healthy option' crops. One example is potatoes with altered starch metabolism which have reduced oil uptake during cooking and, therefore, offer a healthy alternative to the traditional potato chip.

Concerns Surroundings GM Crops

The concept, let alone commercial reality, of genetically modified (GM) crops continues to be the cause of considerable concern to the public. Amongst the most frequently quoted concerns are the fears of gene escape to wild relatives leading to what has been termed gene pollution, the contamination of organically grown crops and the breakdown of disease/pest resistance in GM varieties. Other concerns relate to the safety of ingestion of GM ingredients by humans, for example the potential for developing allergenicity in a crop which was otherwise allergen free. GM ingredients containing antibiotic resistance genes have also met with public resistance due to fears associated with the pootential for antibiotic resistant strains of bacteria developing via gene transfer in the gut of animals or even humans. Biotechnology companies and research organizations are responding to public pressure by developing GM varieties that no longer contain antibiotic resistance markers.

There is no doubt that the recent wave of public concern surrounding the safety and ethics of GM crops has overshadowed the significant potential benefits that GM technology has to offer for those involved in all parts of the food chain from primary producers/ growers to the household consumer,. as well as the potential environmental benefits associated with decreased spray applications. However, it is true that many GM crops may have impacts, including some very positive, upon agriculture and the environment while in some cases there may also be implications concerning food quality and safety. Ultimately there may also be ethical concerns for some sectors of society. Where there are benefits to be gained at a known or unknown risks, the question of risk assessment and subsequent risk magagement arises. If the potential GM technology is to be realised, the quality, safety, benefits and ethical integrity of this new technology must be evaluated against the risks. Where the benefits are found to outweigh the risks the potential of transgenic technology needs to be realised by management under strict regulatory procedures and effective stewardship post-market release.

RISK MANAGEMENT AND AVOIDANCE:

Principles of Risk Assessment

As previously mentioned, GM crops have a number of potential benefits for growers, processors and eventually the consumer, but it is also recognized that are likely to be environmental impacts and implications for food quality and safety. For example, exploitation of novel GM pest-, disease-and herbicide- resistant crops will require different (often reduced) pesticide and herbicide applications. These modified management systems will have an impact upon current agricultural systems and the agricultural environment. Such impacts are best analysed by risk assessments.

The basic concepts of risk assessment for genetically modified crops are similar to those applied to chemical pesticides where the risk is equal to the frequency and the hazard. For example no exposure (frequency) would equate to zero hazard. Risk assessments study both the severity and extent of the hazard or damage as well as the likelihood and frequency at which the damage will occur. Risk is defined as:

$$\text{Risk (impact)} = \text{Frequency (exposure)} \times \text{Hazard}$$

Clearly the ideal situation would be one of zero risk. Since in reality the likelihood of risk is always greater than zero, acceptable risk levels for GM crops must be defined, as with all new technology. What is defined as acceptable is based upon cultural values and may well differ globally. Indeed the current climate of controversy surrounding .GM crops signifies strong cultural differences between European and North American consumers in what is defined as acceptable levels of risk for the utilization of GM crops.

While there are differences in the regulatory procedures controlling the development and commercialisation of GM crops in North Americal and Europe, both systems apply the same broad principles to assessing the safety of GM crop usage for food, animal feed and in terms of environmental impact. The first step involves thoroughly assessing the procedure for modifying the plant tissue. In the UK, for example, the Advisory committee on Genetic Modification (ACGM) is the regulatory authority responsible for contained use evaluation; that is, the initial experimental work 'contained' within the laboratory or glasshouse. The risk evaluation procedure must be specific to each product.Broadly drawn conclusions, for example based on inter-species comparisons, are unacceptable. Most importantly the information requested in a risk assessment must be derived scientifically, with experiments designed to provide clear, interpretable, unequivocal and reproducible results. A recent addition to the risk-assessment procedure has occurred in the UK, in response to public pressure, where there is now a move towards assessing the societal and cultural impacts of this new technology alongside the environmental and human health risks.

Risk assessment can be divided into four steps.

1. problem formulation
2. risk analysis
3. risk characterization
4. risk management.

Problem formulation requires that all available information concerning the plant, the trait and the experimental information is gathered in the context of the most likely hazards, such as toxicity/allergenicity. Once all the data are available, they can be analysed for characterization of the likelihood and/or severity of the risk. In the final phase of the assessment procedure, the acceptability or otherwise of the identified risk must be determined and effective plans set out for its management. The risk assessment procedure is an iterative one and must continue throughout the use of the product, including post-market monitoring.

In the case GM crops there are a number of variables/risk types to consider including impacts on the agricultural environment, closely related species, insects and animals and human health. To analyse the consequences of GM crop impact upon the agricultural environment requires a detailed understanding of the characteristics of the GM crop in question. This involves determining which wild relative, if any, it may hybridize with and studying the management systems involved in growing the GM crop itself. It also involves recognizing any potential effects on other GM or non-GM crops which are likely to be grown in rotation with the variety being assessed. As an example, GM herbicide tolerant (HT) crops will be treated with different herbicides, with different activity spectra, at different crop development stages, leading to effects on the botanical diversity in the GM-HT crop which are the product of the interaction between the GM crop and the herbicide treatment.

The nature of any hazard is dependent upon the characteristics of both the crop that is modified and of the GM trait. Risk assessments require measurement and study of the hazard or impact of both. Numerous studies have concentrated on measuring frequency phenomena such as gene flow and inter-specific hybridization without considering the impact of the transgene when it has dispersed or introgressed into other population or species. In addition the impact of the release of the GM plant will depend on the type and location of the environment into which it is being released. To be truly effctive, risk assessments may have to be carried out for a range of locations as they are not necessarily transferable from one site, area, region or country to another.

Impact of Plant Species

Plants vary in the degree to which they are dominant or are invasive in certain environment and in their ability to disperse genes to different populations and species. They will therefore have different environmental impacts when genetically modified. For any particularly country or region, plants can be classified as potentially being high, medium or low impact.

Plants in the high-impact group are generally hardy, perennial, competitive, open-pollinating and prolific having a wide range of relatives with which they hybridize and an ability to colonise a range of natural and semi-natural habitats. Examples include perennial rye grasses (*Lolium perenne*) and certain indigenous and introduced trees and shrubs that form a significant proportion of forests and woodlands, e.g. *Populus* spp. Modifications of these plant, which affect their competitiveness, could have significant impacts upon the ecology of a range of environments.

Medium-impact plants are open-pollinating, hybridize with some wild relatives, are prolific and colonise a limited range of habitats. Examples of such plants include oilseed rape, oats, sugar beet and rice, all of which have closely related wild relatives with which they hybridize and an ability to colonise disturbed ground. These plants and their close relatives rarely form climax populations except in particular environments such as coastal areas or in disturbed ground. Low-impact plants are usually annual or biennial species, are largely self-pollinating with few hybridizing relatives that are poorly adapted (or not native) to the area in which they are cultivated. In the UK, examples include maize and sunflower.

It is important to appreciate that the impact of a plant species, will depend upon the environment into which it is being released. Maize and potato are considered low-impact plants in England. However in Central and South America, where their centres of genetic diversity occur, their impact would be considered very high.

Impact of Transgenes

Transgene expression in GM plants will have different impacts in different environments. Since genes often operate uniquely it is not easy to classify transgenes as having high or low impact. In addition their impact is also dependent upon the nature of the receiving environment (Agricultural impact).

High-impact transgenes generally encode genetic modifications that improve the fitness of the GM plants by increasing their reproduction, comptitiveness, invasiveness and/or persistence and will therefore also have the greatest environmental impact. Thus transformations that significantly increase plant productivity by overcoming constraints such as broad-spectrum pest, disease and stress tolerance will have the highest impact. Many pest- and disease-resistance genes will have effects on non-target species either directly or indirectly by altering relationships between pests and beneficial organisms. It is important that these non-target effects are thoroughly understood before commercialisation progresses.

Low-impact transgenes are genes that do not noticeably enhance the fitness of the modified plant so that the modified plant's role and behaviour in a given ecosystem is not altered. Examples would include genes that modify seed composition, e.g. high lauric acid genes in oilseed rape and high starch content genes in potato. However, in preparing a comprehensive risk assessment it would be important to confirm that low-impact genes might not, unnintentionally, confer an environmental advantage. As an example, in the case of high starch content genes in potato, it would be important to assess that the transgenes do not significantly increase potato seed tuber over-wintering survival rates through enhanced frost resistance. In the case of oilseed rape, it would be important to ensure, for example, that there is no increase in the dormancy characteristics of oilseed rape which may confer enhanced soil survival characteristics.

Mechanisms of Transgenes Transmission

Gene flow is an important consideration in evaluating the risks associated with growing GM crops. Transgene dispersal could lead to contamination of neighbouring crops, a particularly worry since the UK organic authority amended its rules to include a zero tolerance to the presence of GM material. Transgene flow from crops to closely related wild relatives is also of concern as an environmental risk. Gene flow between different species is, however, not a new concept and has in fact been occurring between natural plant species, leading to a range of hybrids in the UK flora including amongst others the Salix, Lolium and Rumex genera.

In order for gene transfer from one species of plant to another closely related wild relative to occur a number of barriers, both physical and genetic, must be overcome. These include dispersal (either of pollen or seed), longevity of the pollen grain,: sexual compatibility, competition with other pollen sources and events post-fertilization. Most gene dispersal occurs as a result of pollen transported either on the wind or via vectors such as bees or, less commonly, by seed dispersal. The distances over which pollen dispersal occurs varies depending upon the plant species, the prevailing weather conditions, in the case of wind-borne pollen, or the insect vector. As discussed by Moyes and Dale (1999) although most studies have concentrated on the range of pollen dispersal, the survivability over times of the pollen grain is actually the most important aspect of potential gene transfer and cross-contamination.

Assuming that pollination is successful and gene transfer has occurred, the barriers to successful introgression of a gene from the original donor species to the recipient will be dependent upon what the gene might offer the recipient. If, for example, the gene induces a lethal effect, the seed of the recipient plant will die and gene introgression into the recipient species will go on further. If, however, the transgene confers a selective advantage such as cold tolerance, drought or disease resistance or the ability to thrive in low-light conditions, seed from the recipient plant will thrive. This is especially true for native species, with the greatest opportunity for transgene movement occuring within the crop-weed complex. However, in assessing the scale of transgene movement, it is important to consider whether those plants containing genes conferring an adaptive advantage in the agricultural environment might lose that selective advantage in the differing environmental conditions outside of the farm field. If the transgene provides no selective advantage to the recipient plant, such as herbicide-resistance genes present in plants growing in an environment where herbicide spraying will not occur, the transgene will have a neutral impact upon the recipient species. There will be no incerease in fitness of the population.

Multiple Transgenes and Transgene Stability

One of the major issues surrounding GM crops containing multiple transgenes encoding a variety of traits is the question of stability of gene expression. Might the introduction of a second transgene affect expression of the original transgene and thus the phenotype of the GM variety? In particular, genetic homology between the two transgenes may cause down regulations of gene expression and suppression of the phenotype. How this effect is caused is complex and thought to be affected by factors such as the position of the transgene within the genome, i.e. point of insertion during the transformation procedure, transgene copy number within the genome and by other factors such as reproduction and even environmental conditions. The results may be unpredictable resulting in instability or silencing of gene expression. The production of GM varieties, involves evaluation of transgenic lines over a number of generations, during which any unstable lines would generally be identified and discarded. One possible exception to this would be instability arising from environmental interaction. This instability is also observed in conventionally bred varieties, providing a basis for further analysis of GM varieties.

From the prospective of risk assessment and environmental impact the most significant issue arises from gene flow between closely related species. Instability of gene expression generally leads to suppression of gene expression, in which case the phenotype of the GM variety would revert to the wild type, with no expression of the transgene. The implications for agronomic practice are significant, as suppression of gene expression would render a herbicide-tolerant GM variety susceptible to that particular herbicide, with consequent loss of yield if the farmer were to spray unwittingly. While the effect of transgene instability on the natural environment is likely to be minimal, there may be important issues at stake in the case of transgenic plants engineered to remove the synthesis of harmful toxins. In this situation suppression of gene expression arising from gene flow leading to multiple transgene insertions could prove a serious human or animal health problem if undetected.

IMPACT OF GENETICALLY MODIFIED CROPS

Impact on Agricultural Systems

Genetic modification can have a range of impacts on agricultural systems and therefore will require specific agronomic management. The use of GM varieties would affect the nature of crop volunteers in subsequent crops and require alterations in volunteer management practices. The GM trait may also have an impact if it disperses to other crops and weeds through cross-pollination and seed dispersal. Low-impact genes such as herbicide tolerance, which have little impact of natural environments, become highly significant because of the changes in the herbicide usage required for their management. These herbicides will differ in the effect they have on plant and other species diversity in cropped fields.

Deployment of high-impact genes such as those encoding pest and disease resistance will result in reductions and changes in pesticide usage and thus offer opportunities to enhance diversity in cropped fields, especially if the transgene products are very specific to selected pests. However, it is imporant that the selection pressures they impose on pests and diseases do not encourage the development of virulent races of pests and pathogens and appropriate management systems are required in order to maintain durable resistance in the GM varieties.

Impact of Uncultivated Flora

Genetically modified crops may also have impacts on uncultivated and 'natural environments. These environments may be affected by characteristics of crop and wild species induced by novel genetic constructs and their products. Risk assessments must therefore concentrate on whether the genetically modified characteristics of a GM crop and of similarly modified hybridizing wild relatives are likely to change the behaviour of the plants or dependent flora and fauna in their environment, to the extent that ecological balances are altered.

Impact on Insects and Animals

The first successful example of using a foreign plant gene to confer resistance to insects was reported in 1987 and involved transformation of tobacco (*Nicotianum tobacum*) with the cow pea trypsin inhibitor (CpTi) gene. Since then there have been many reports of success in insect management using transgenic crop varieties.

The bacterial endotoxins isolated from *Bacillus thuringiensis* (*Bt*), comprise one of several groups of proteins which have been shown to have insecticidal properties to a range of economically important insects. Transgenic crop varieties engineered with *Bt* resistance are already in commercial use in the USA and China, while a number of plant proteins, such as inhibitors of proteases, lectins and other digestive enzymes, are being evaluated for their efficacy as insect-resistance mechanisms.

It is important that genes selected for the control of insect pests have acceptably little effect on non-target insects including predators of the target pest insects, in order to maintain insect diversity in GM crops. Clearly, if there is an effect upon predators that is comparable with current control practices then little benefit will accrue from the deployment of GM crops, a point made strongly by the Royal Society for the Protection of Birds in their submissions on GMOs (1997). Impact assessments are therefore required to examine the effects on non-target organisms in the crop enviroment.

Studies to evaluate the effect of transgenic plants expressing insect resistance on non-target species have provided, at best, equivocal and often controversial results which have served only to fuel the GM debate rather than provide hard scientific facts on which to base a thorough impact assessment.

Research into the impact of potato plants expressing the snowdrop lectin GNA upon 2-spot ladybirds which feed on the arphid *Mysus persicae* demonstrated that the ladybirds were affected adversely in terms of fecundity, egg viability and longevity. However, the authors point out that the effects may either.by a direct result of ladybirds preying on aphids which have digested transgenic plant material containing the lectin, or may also be due to poor nutritional quality of the aphids themselves as a food source. Other studies involving the parasitic wasp *Eulophus pennicornis* and the tomato moth *Lacanobiua oleracea* demonstrated that the parasitic wasp was not affected when it parasitized moth larvae reared on transgenic potato plants expressing the snowdrop lectin GNA.

More recently published a report indicating that pollen from transgenic *Bt*-resistant maize plants had a detrimental effect on the larvae of the non-target Monarch butterfly (*Danaus plexippus*), which is considered to be a sensitive indicator of environmental disturbance in the USA. Larvae, which normally feed on the leaves of the milkweed (*Asclepias curassavica*) plant were fed on leaves that had been dusted with unqunatified amounts of pollen from the transgenic *Bt* maize plants. Results indicated that larval survival rate was only 56% compared to 100% survival for larvae fed on leaves dusted with untransformed pollen. Superficially these results indicate an unacceptable environmental impact from *Bt* maize. However, closer analyses have revealed a number of serious criticisms of the research, including the use of laboratory studies only, no-choice feeding regimes, lack of stringency, lack of quantification and the use of inappropriate controls. The experiments were not conducted in the field so no *in vivo* data were available to confirm that (A) milkweeds occur in maize fields, and (b) that Monarch butterflies occur on these milkweeds bearing in mind the insecticide programme received by conventional maize. This once again reiterates the requirement for comprehensive risk assessments based on thorough science.

Schuler *et al.*(1999) have conducted research concerning the environmental effects of *Bt*-resistant GM oilseed rape on a non-target insect. The results demonstrated that the behaviour of non-target insects can also play a part in determining how *Bt* plants will affect their populations and should be considered when trying to evaluate the environmental impact of GM crops. Their laboratory-based experiments evaluated the ecological impact of the GM crop on the diamondback moth (*Plutella xylostella*), a pest that damages the oilseed rape crop, as well as the natural bio-chemical agent of the diamondback moth, a parasitic wasp (*Cotesia plutellae*), which kills the moths' caterpillars by laying its eggs in them. Results demonstrated that parasitoid wasp larvae that were oviposited in *Bt*-suspecptible moth larvae not surprisingly died with their hosts,. In contrast wasp larvae that had been oviposited in *Bt*-resistant moth larvae feeding on transgenic plants survived and demonstrated no adverse effects of exposure to the *Bt* toxins either as adults or in the development of their own larvae.

The research group then examined the behaviour of the female parasitic wasps in the presence of GM and non-GM leaves. It is known that the female wasps locate the host diamondback moth larvae using herbivore-induced volatiles released from the damaged plants. A wind-tunnel was used to compare the flight response of the wasp towards *Bt*-susceptible and *Bt*-resistant diamondback larvae which were allowed to feed on *Bt* leaves. The flight and feeding behaviour of each wasp was then measured. In this test, 79% of the parasitoids flew to the *Bt* leaves damaged by resistant moth larvae, with only 21% choosing *Bt* leaves damaged by susceptible larvae. The apparent lact of effect on the survival or host-seeking ability of the parasitic wasp suggested that *Bt* plants may have an environmental advantage over broad-spectrum insecticides.

Impact on Human Health

The potential for transferring genes from one unrelated species to another has caused concern that allergenicity may be introduced into a food source that was previously non-allergenic. An obvious example is that of the recent research programme by Pioneer Seeds where soybean transformed with a gene from the Brazil nut was found to have allergenic properties. The research programme was halted before any field trials took place but public concern was heightened by reports of this work. All GM foods are now routinely tested for allergenicity using serological test involving immuno-globulin antigens for specific allergenic proteins.

Transgene instability may be an important issue in the case of transgenic plants engineered to remove the synthesis of harmful toxins. In this situation suppression of gene expression arising from gene flow leading to multiple transgene insertions could prove a serious human or animal health problem if undetected.

The inappropriate choice of transgenes for achieving a desired trait may also have a serious impact on human health without adequate risk assessment. Lectins are a group of proteins known to have insecticidal properties that make them attractive candidates for the development of transgenic plants with resistance to the Homoptera insects. They are thought to work by binding carbohydrate side chains present in the gut wall resulting in inhibition of food absorption. As there is the potential of toxicity to humans, it is essential that extensive risk evaluation is required to established and potential threats of toxicity. The need for such risk assessment is reflected in work on GM tomatoes and was recently highlighted by reports from the Rowett Institute in Scotland, which indicated adverse immunological and nutritional effects from enhanced lecithin in GM potatoes. However, aspects of the former of these reports in particular were strongly criticized by a number of reputable scientific bodies as being unsubstantiated and have highlighted the need for agreed methodology in this field of research so that conclusive results can be acquired.

There has also been concern about genetically modified ingredients containing antibiotic resistance genes used to select transformed cells prior to the regeneration of transgenic plants. Use of these genes raises the potential for antibiotic-resistant strains of bacteria to develop via horizontal gene transfer in the gastro-intertinal tract of animals or even humans. This possibility is not thought to be a major hazard since the antibiotic-resistnace genes most often used for plant transformation themselves come from bacteria. They encode resistance against antibiotics rarely used in medicines such as kanamycin, against which a large percentage of gut microflora is already resistant.

However, given the risk, however small, of producing more antibiotic-resistant bacteria, techniques are being developed that will enable selectable markers to be removed from crop plants after the transformation process. Alternatives selectable markers, not based on antibiotic selection, are also being tested, for example a mannose permease that allows the use of mannose, a sugar not normally available to plant metabolism, as a carbon source during plant regeneration.

Risk assessment methodology will also have to be adjusted for food plants which are modified to improve nutritional and other qualities, a major area for current and future research. Target traits include, for example, improving the nutritional value of proteins, increasing the concentrations of oils low in saturated fats, or fortification with micronutrients or antioxidants. Food plants modified in this was must undergo extensive toxicological and nutritional assessment with a combination of *in-vitro* and *in-vivo* tests, as currently required for all novel foods by the EU, for example. In the case of genetic modification, however, particular attention needs to be given to the detection and characterization of potential unintended effects of modification. Inferences about such effects can no longer be based solely on chemical analysis of single macronutrients and micronutrients, and known crop-specific antinutrients or toxins. New methods have been developed to screen for potential alterations in the metabolism of the modified organism by such methods as:

- analysis of gene expression (monitored, for example, by microarray technology or mRNA fingerprinting)
- overall protein analysis (proteomics)
- secondary metabolite profiling.

Studies using these designs will need to be designed carefully to take account of the complexity of foods.

HOW IS BIOTECHNOLOGY REGULATED?

International Picture

An an international level, a joint FAO/WHO initiative has agreed a basic framework for evaluating the safety of transgenic food crops. Key aspects of any evaluation should include:

- characterization of the new gene product
- identification of alterations in concentrations of nutrients and known toxicants
- assessment of the potential allergenicity of the gene product
- assessment of the impact of gene transfer between plants and the gut microflora of animals and humans.

An early basis for evaluation has been the concept of 'substantial equivalence'; how a new food compares with a conventional product that experience and use has proved to be safe for consumption.

Although there is some consensus on approach, there are as yet no internationally agreed procedures or advisory bodies in place to regulate the release, either experimentally or for commercialization, of GM crops. The approach to the regulation concerning releases to the environment differs between countries, although at the time of writing discussions concerning a global regulatory system are under way. The one exception is the European Union which has co-operated to form an agreed framework of legislation through EC Directive 90/220 *on the deliberate release into the environment of genetically modified organisms*. This directive, while ultimately seeking a unanimous decision concerning consent to release for commercialization, leaves the individual member state free to decide whether to grant permission for release for experimental purposes.

UK Regulatory Process

For a new GM variety to be released into the environment in the UK, and application for consent to release, which include a risk assessment, must be made. Interestingly, the UK Competent Authority, the Department of the Environment, Transport and the Regions (DETR) has decided to adopt a policy of transparency in these evaluations. As such, all information on every application for consent to release (except commercially sensitive data), is placed on the Public Register in summary form and is available in more detail on application to the DETR. In addition any application for release for experimental or commercial purposes, without exception, must be advertised publicity in the vicnity of the release site (normally a local newspaper) and the consent approval, conditions of release and location of the release are made available via the Public Register.

Process in details

In the UK responsibility for the safe release and subsequent monitoring of GM crops lies with the Department of the Environment, Transport and the Regions (DETR), which is recognized within the EU as the UK's Competent Authority. The DETR as the lead department, in consultation with all the other relevant UK departments, has outlined very specifically the requirements for anyone wishing to grow and market GM varieties in the UK. The applicant must notify the Health and Safety

Executive (HSE) of their intention to initiate research work at the laboratory stage and produce an assessment of any risks to be environment or human health. The HSE will then consult the Advisory Committee on Genetic Modification (ACGM). The ACGM is responsible for regulating the contained use (laboratory or glasshouse) of Genetically Modified Organisms under the Genetically Madified Organisms (Contained Use) Regulations 1996 which came into force on 27 April 1996. Along with European Directive 90/219 the regulations establish a notification scheme for activities with genetically modified organisms (GMOs) according to their potential harmfulness to people or to the environment.

If the variety is to be grown outside for field trials a detailed application and environmental risk assessment must be completed by the applicant and submitted to the DETR. Applications for consent to release a GM variety into the environment are evaluated depending upon the nature of the release application, i.e. whether the request is for release for experimental and evaluation purposes (Part B release) or the applicant is requesting permission to release the GM variety for commercialization (Part C release).

The risk assessment needs to contain information concerning the nature of the inserted transgene(s) and a detailed characterization of the genetically modified plant arising from transformation with the transgene(S). The following information is sought:

- source of the transgene(s) and additional sequenc
- details of the modification
- method of transformation
- level of interaction
- level of transgene(s) expression
- environmental interaction
- impact on human health
- management considerations for the relases site.

The DETR is advised by the Advisory Committee on Releases to the Environment (ACRE), a statutory committee responsible for considering all Part B releases (release for experimental purposes only). A second committee, the Advisory Committee on Novel Food Processes (ACNFP), responsible to MAFF, advises on food safety issues. There is yet another committee, also answerable to MAFF, which is responsible for considering issues surrounding the use of GM varieties in animal feed. In evaluating GM issues all three committees may be required to provide advice and guidance.

ACRE's judgement on whether or not to grant consent for release for experimental purposes is based on evaluating the impact of answers to the following: is it possible that the inserted gene(S) may

- make the modified crop more persistent
- make the modified crop more invasive
- make the modified crop more undesirable to living organisms or the environment
- be transferred to other organisms?

In an application is successful a consent from the Secretary of State for the Environment will be given allowing the varieties to be grown outside for experimental purposes (Part B consent).

If the variety is progressed from experimental release to commercial field trials then the ACNFP would consider the new variety for food safety purposes under the EC Novel Foods regulation. The ACNFP holds responsibility for evaluating the safety of any novel foods submitted for approval under the EC Novel Foods and Novel Food Ingredients Regulation (258/97). This regulation, which came into effect on 15 May 1997, defines a novel food as food that has not been used for human consumption to a singificant degree within the Community. What this means in reality is that foods which have been sold in one or more member states before 15 May 1997 will not generally be covered by the regulation. This includes food ingredients obtained from the herbicide tolerant GM soybeans and insect-resistant maize, which were approved for sale under the Deliberate Release Directive 90/220/EEC. These food ingredients, which are on sale in a variety of processed foods, are not regarded as novel foods but are covered under EC regulation 1139/98 which requires the labelling of food ingredients containing GM material derived from soya or maize.

The ACNFP may then seek advice from the Food Advisory Committee (FAC), Committee on Toxicity of Chemicals (COT) and the Committee on Medical Aspects of Food Policy (COMA). A marketing consent is also required before the crops can be grown commercially, which requires another detailed submission to the DETR for consideration by ACRE. Once these committees approve an application their assessment is passed to other EU member states for consideration. Finally, if the new GM variety has been awarded Part C consent by the Department of the Environment, Transport and the Regions (DETR) the new variety may go forward for statutory evaluation for plant variety registration in the member state with a view to commercialization. If no objections are raised and the new variety performs well, the product could be marketed, provided it is labelled in compliance with the new EU food labelling laws.

In the UK any new crop variety, including genetically modified varieties, targeted for commercialization must comply with the regulatory requirements under the UK seeds legislation, the UK Environmental Protection Act 1990 and the GMO (Deliberate Release) Regulations 1992 and 1995, in addition to the previously mentioned EC Directive 90/220.

GM and non-GM varieties must also undergo statutory evaluation for registration on the National List, leading to plant breeder's rights and eventually the European Common Catalogue. This requires evaluation of the Distinctiness, Uniformity and Stability (DUS) of a variety and its Value of Cultivation and Use (VCU). A GM variety will not be added to the UK National List until it receives full part C consent from the EC, covering all issues relating to the environment, human health, animal feed and human food. In other words, full Part C consent is the prerequisite for variety registration and commercialization. Once the new GM variety has successfully passed through all these evaluation phases a breeder may consider commercial development of the new policy.

The current regulatory system in the EU based on Directive 90/220/EEC has been heavily criticized due to the length of time taken for applications for consent to release to be evaluated. The delays arise as a result of the evaluation process, which looks at each application, based on the process (how the GM crop is derived) rather than the end product (modified trait). The product-orientated approach operated in the USA and Canada results in a much faster regulatory procedure. Discussions are currently under way aimed at reviewing the EU procedures.

Regulation in the USA and Canada

The USA and canada have adopted a different approach to the regulatory process from that taken by the EC where the regulations relate to the process of modifications. In the USA the regulation relates to the product and not process, while in the Canada the regulation relates to any plant with novel traits, which may include plants developed by approaches other than genetic modification.

The product-driven approach in North America means that once a product is proven safe it is no longer regulated. Similar releases by other organizations employing the same crop species/trait modification are subject to notification only, resulting in a fast track for proven modifications. The criteria used to determine safety to the environment and human health are equally rigorous to those considered in Europe, but the product-driven approach leads to faster approval and commercialization.

The agencies responsible for regulating biotechnology in the USA are the US Departement of Argiculture (USDA), the Environmental Protection Agency (EPA) and the Food and Drug Administration (FDA). The USDA has responsibility for regulations concerning plant pests and plants. Within the USDA the Animal and Plant Health Inspection Service (APHIS) is responsible for protecting US agriculture from pests and diseases. Notification is required for the introduction of a genetically modified organism that is considered to be a plant pest, such as a modified plant with weedy characters or a modified pathogen.

The EPA is responsible for regulating novel micro-organisms, microbial and plant pesticides and new uses of existing pesticides including use on GM herbicide tolerant crop. The FDA has regulatory responsibility for food, feed, and food additives derived from new plant varieties. The FDA policy requires that genetically modified ingredients meet the same safety standards as required for all other foods. Products are regulated according to their intended use, which means that some products will fall under more than one regulatory agency. For example, viral resistance in a food crop would be evaluated for safety to grow by the USDA, environmental safety by the EPA and by the FDA for safety to human health.

Food Labelling

Novel foods such as GM foodstuffs are, where feasible, assessed in comparison with the foods they will replace. The process of substantial equivalence, as this assessment is known, was developed by the World Health orgnization (WHO) and the Organization for Economic Co-operation and Development (OECD) and is an internationally accepted evaluation procedure. Evaluation of a GM foodstuff involves a wide range of information including agronomic aspects of the plant material and detailed information concerning the nutritional composition of the material. The evaluation looks at the intentional effects of the modifications as well as any unintentional effects.

In addition to standard labelling requirements for food, the EC Novel Foods and Novel Food Ingredients Regulations outlines specific labelling requirements intending the consumer to be notified of:

- Any characteristic or food property which renders a novel food no longer equivalent (practically interpreted as meaning where GM material, either DNA or protein, is present in the food or ingredient) to an existing food or food ingredient.
- The presence in the novel food or food ingredient of material which is not present in an existing equivalent foodstuff and which may have implications for the health

of certain section of the population or give rise to ethical concerns. Examples would include allergenic potential or the use of animal concerns, which may be the cause of religious concerns.

- The presence of an organism genetically modified by techniques of genetic modifications as defined by MAFF guidelines.

The Food Labelling (Amendment) Regulations 1999 came into force in March 1999 in the UK and require that catering outlets, without exception, provide information concerning GM ingredients for their customers. The regulations apply to food and food ingredients produced in whole or part from GM soya or maize, but exclude foods where neither the protein nor DNA resulting from the genetic modification is present. In these cases the food is deemed indistinguishable from products produced using conventional soya or maize. Food additives, flavourings for use in foodstuffs or extraction solvents are also exempt as are ingredients lawfully sold before 1 September 1999. Similarly, products placed on the market before 1 March 1999, where other forms of wording (complying with EC Regulation 1813/97) have been used to indicate the presence of GM material, are also exempt. Further details can be obtained for Guidance Notes: Novel Foods and Novel Food Ingredients Legislation, and Guidance Notes: Labelling of Food Containing Genetically Modified Soya and Maize, MAFF.

PUBLIC PERCEPTIONS

Opponents of genetic engineering are concerned that the technology is unproven and that not enough is known about the potential risks to human health or the environment. They argue that once genetically modified plants are released into the environment it will be very difficult to repair any damage caused by their release, and that humans are being used in a live global experiment to evaluate the safety of GM ingredients.

Proponents of genetic engineering argue that the application of genetic engineering is controlled by strict regulations concerning the release of GM varieties into the environment. The regulations are concerned with ensuring that deliberate release, either for experimental or commercial purposes, does not pose a risk to the environment or to human health.

American Versus European Response

Despite the complex evaluations that a GM variety must pass before commercialization becomes a possibility, coupled with transparency of the procedures, the British public has become increasingly worried by the development of these new varieties. Greenpeace commissioned a MORI poll in June 1999 to evaluate public opinion regarding the potential risk of gene transfer from GM crops to neighbouring organic crops. Mori approached 501 adults over the age of 16 and asked the following question:

As it is currently agreed that some cross-pollination of GM crops with neighbouring organic crops is inevitable, it has been proposed that the standard for defining 'organic' foods should include organic crops which may have cross-pollinated with GM crops. How concerned, if at all, would you be if the definition of organic crops was changed in this way?

The results of the poll indicated that 45% of those questioned were very concerned, 29% fairly concerned, 15% were not very concerned, 8% were not at all concerned while 3% either did not know or had no opinion. The concern of the British public is in contrast to the attitude of North American consumers who have, until now, largely accepted GM foods.

The reasons for the difference in attitude between American consumers and European, in particular British, consumers have been debated long and hard, with the conclusion that differences in public perceptions are likely to be attributable to a number of causes.

- *Trust in Government and statutory bodies*. A catalogue of poorly handled crises in the British food industry, ranging from the salmonella scare in eggs of the mid-1980s through to the recent BSE crisis, have left the public sceptical and mistrusting of official attempts to reassure it about the safety of GM technology. This distrust is in spite of the clear and significant differences in the actual risks associated with

GM foods compared with these earlier incidents. In contrast, the USA has not experienced this crisis of confidence in the regulatory authorities and when the FDA move to reassure the public they appear to be successful.

- *Awareness of GN foods and what genetic modification involves.* Recently it has stated to become clear that the British public are considerably more aware of what genetically modified food is then their American counterparts. When consumers in the USA were interviewed at supermarkets to find out how aware they were of what GM foods are and how they regarded them, a significant proportion of those interviewed did not know what a GM food was and had not considered whether they may be desirable as part of their daily diet. The British environmental groups are beginning to report an upsurge of public concern in the USA in response to the furore seen in the UK and the rest of Europe.
- *Differing attitudes to the rural environment.* There are clear difference in attitude to the countryside in the USA principally due to the way food crops are farmed. While crop production in he UK and the rest of Europe forms an intrinsic part of our countryside and indeed has been largely reponsible for creating the present look and feel of our countryside, crop production in the USA tends to occur at arm's length from the consumer in large acreages of land devoted to nothing else but highly intensive agriculture. In that sense there is a perception that the risk of gene transfer and environmental impact of GM crops on American wild-life is minimized and almost preventable whilst in the UK agricultural system the risk of environmental impact would be difficult to contain should a problem arise.
- *Lack of consumer choice.* Of all the reasons for the current rejection of GM technology by the public, the lack of consumer choice is perhaps the most important. The decision to import GM and non-GM soya without segregation may well be taken as the turning point in

> consumer attitude against these new crops. Recognition that consumers demand the right to choose whether they eat GM foods has orchestrated a remarkable volte face by the large-scale food processors and supermarkets. They have moved almost unanimously to remove GM ingredients from their products. The knock-on effects can be seen in the elevated prices of last season's non-GM cereal and legume crops which commanded a premium.

New food labelling regulations came into force in the UK in 1999 and are aimed at meeting the demand for consumer choice. They require that catering outlets display information indicating that a product does or does not contain GM ingredients, with a maximum penalty of *5,000 for non-compliance. However, it is difficult to see how the scheme will work in practice. Given the technical demands of analysing and identifying GM ingredients, it will prove difficult for the local authorities to enforce.

The scientific community and industry are beginning to recognize that one of the problems in obtaining public acceptance of GM crops is the lack of benefit that these crops currently offer to the consumer. The development of GM crops has so far concentrated on producing resistance-enhanced crops. Herbicide, disease and insect resistances are the most prolifically produced GM crops world-wide. The benefits of these new traits are realised by the primary producer, the farmers, who do not have to spend money and time spraying to prevent crop damage. In the second major category of GM crop development, the emphasis continues to be on the added-value traits such as novel oil production. The benefits are enjoyed by processors rather than consumers.

Finally, in the smallest category are the designer traits such as improved nutritional value, shelf life or reduced cost, which is immediately obvious to the consumer. Even within this category the question of whether trait improvements such as slow ripening can be seen, or effectively sold, as a benefit to the consumer, rather than the wholesalers and supermarkets, is questionable. This lack of demonstrable benefits of genetic engineering to the public does little to diminish the suspicions that the real winners are the biotechnology companies, growers and processors.

FUTURE DEVELOPMENTS IN THE REGULATORY PROCESS

Post-Commercialization Regulatory Systems

In view of the current upsurge of anti-GM feeling the regulatory controls surroundings the deliberate release of GM crops is likely to become even more stringent. Current regulations are under review with the aim of improving efficiency of the current procedures. Proposed changes to Directive 90/220/EEC include:

- Setting clear time limits during consideration of marketing release applications with the aim of giving a decision concerning an application within one year.
- Clarification of the risk assessment and harmonization across member states to include assessment of direct, immediate and delayed risk or impact.
- Monitoring and time limitations consents, such that monitoring will occur post-release for commercial purposes. In addition marketing consents will have a finite life of ten years after which they must be reviewed.
- Improving the transparency of applications for consent to release for marketing, allowing the public to see the content of such applications.
- Appointment of a committee to deal specifically with the question of ethical issued arising from the application of biotechnology.

Risk assessments conducted for regulatory purposes tend to concentrate on the direct effects of the GM crop and its relatives on the natural environment. However, it is now becoming apparent that the agricultural consequences of the deployment and management of GM crops could have significant impact on the environment in regions where a very high proportion of the total land area is actively managed. Plans are being developed for monitoring the early years of the commercialization of each GM crop so that its impact on both agriculture and the environment can be evaluated in farm-scale releases. The aims of such large-scale monitoring programmes are to:

- evalutae any different or unforeseen risks that might be associated with large-scale releases
- try to provide mechanisms of quantifying such risks
- try to understand the interactions of GM traits, for example herbicide tolerance, within the context of standard agricultural practice, e.g. crop rotations, over a longer period
- develop safe codes of practice to minimize risk.

There are a number of potential approaches to monitoring the post-release effects of GM crops. At this juncture in time when no GM variety has been released for commercial growth in the UK, and when the likelihood of imminent release is low, the priority must be to monitor experimental work.

Stewarding and Voluntary Codes of Practice

As a response to customer concerns the UK agricultural and biotechnology industries have come together to form a stewardship group with its own voluntary codes of practice in order to develop an industry-wide code of practice for the commercial introduction of GM crops and their subsequent management. The group, known as the Supply Chain Initative for Modified Agricultural Crops (SCIMAC) represents farmers, the seed trade, plant breeders and the agrochemical and biotechnology companies and was launched in June 1998.

SCIMAC seeks to provide traceability for individual consignments of GM varieties and outlines best practice for the provision of information at successive points along the food chain. Traceability begins with the plant breeders who must comply with EC regulations concerning the release of GM varieties into the environment for commercial purposes as discussed proviously. SCIMAC then encourages seed merchants and growers to label clearly seed of GM varieties and to advise their respective customers of specific requirements concerning the growth, management and handling of these novel crops. Farmers must isolate GM crops from similar non-GM crops to prevent hybridization and must keep all harvested products separate from non-GM crops. Merchants and wholesalers are required to maintain records concerning the storage and subsequent distribution off the farm. Any GM consignment to leave the farm must include a post-harvest declaration stating the variety name.

Food Labelling

At the moment the most likely developments in the regulatory procedures are concerned with further modifications to food labelling requirements. The UK competent authority is expected to continue lobbying for the inclusion of food additives, such as lecithin, under the novel food regulations.

Multiple Transgenes and Risk Assessment

The development of transgenic crops containing more than one transgene is already well under way, with a number of oilseed rape varieties containing herbicide tolerance genes and transgenes creating male sterility and restoring male fertility in hybrid varieties. These multiple transgene GM crops further complicate the risk assessment and may influence the decision process when considering applications for consent to release into the environment.

6

Genetically Improved Food

Many of the initial studies in plant biotechnology were focused on developing new plant varieties with better yield rather than changing the properties of the plant-derived fruits. Thus, genes conferring resistance to several biotic and abiotic stresses were incorporated in cultivated plants. As a consequence, since 1995 a number of crops with improved agronomic traits have been available for farmers in some parts of the world. Biotechnology has also the ability to create new varieties focused on product quality and output traits rather than agronomic traits. Fruit commercialization of a genetically engineered clonal plant with modified oil content occurred in 1996. It was a milestone in the long path followed by many research groups towards genetically modified plants with improved properties for human consumption. This has been partially the result of combating the attitudes of consumers to genetically modified foods. There aspects constitute the topics reviewed in the present chapter. In addition, it has recently become clear that the use of transgenic plants as living reactors constitutes an advantage for the inexpensive production of some proteins and metabolites that are economically important. This new trend in the application of plant biotechnology is also reviewed in this chapter.

CHANGING THE NUTRITIONAL VALUE OF FOODS

Amino Acid Content of Proteins

An early application of biotechnology for improving the nutritional value of foods has involved changing the amino acid composition of some common proteins of the human diet. It has been long known that humans cannot live on a protein-free diet. The reason is that we are incapable of synthesizing half of the 20 standard amino acids present in proteins. These are known as essential amino acids and must be provided in the diet.

Consequently, the nutritive value of a protein-based diet is directly related to the content of these essential amino acids. In general, cereals have proteins with a low content of the essential amino acids lysine, threonine and tryptophan, while legume proteins have a deficiency of cystein, methionine and triptophane. Among the most valuable sources of vegetable protein are the grain legumes. In one of these, soybean, genetic engineering has been applied to increase the content of the essential amino acid lysine in the seed proteins. The rationale was that increasing the synthesis of lysine in the seeds of soybean would increase the synthesis of proteins with high lysine content. Lysine synthesis in this species is finely regulated by a feed-back mechanism, i.e., when the lysine content is high there is an inhibition of two of the enzymes involved in the metabolic pathway leading to the synthesis of this amino acid.

The strategy consisted of the integration in the genome of soybean of genes from other species encoding for enzymes without the feed-back mechanism. The transformation of soybean with the gene *lysCM4* from *Escherichia coli* (encoding AK) and the *dapA* gene from *Corynebacterium* (encoding DHDPS), both insensitive to the feed-back inhibition by lysine, resulted in a transgenic soybean plant with a duplicated pathway of lysine biosynthesis, one sensitive and the other insensitive to lysine. As a result, the lysine content of the transgenic soybean plants was over 100-fold the value of the untransformed plants. Other plant species like corn, wheat and canola have been subjected to the same genetic manipulation to increase their lysine content with similar results to those obtained in soybean.

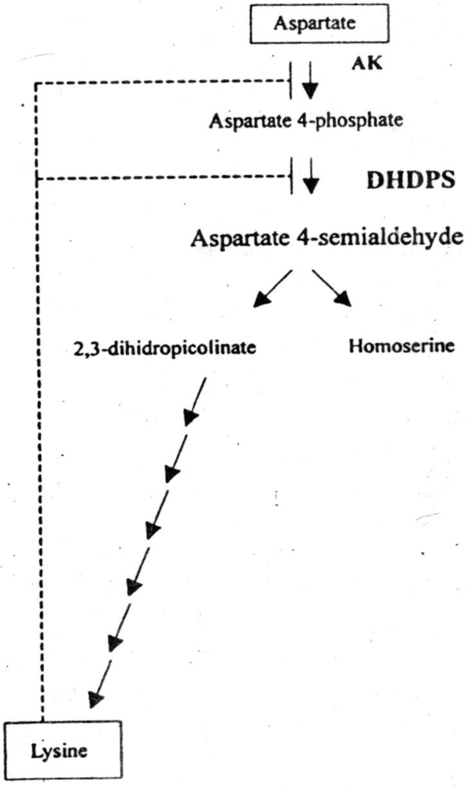

Figure. 6.1: ***Biochemical regulatory mechanism proposed to regulate the synthesis of the amino acid Lysine, in higher plants, derived from the amino acid Aspartate. Enzyme activities. Aspartate kinase (AK) and dihidrodipicolinate synthase (DHDPS) are indicated. Broken lines indicate the inhibition of these enzymes by Lysine.***

Fatty Acid Composition of Triacylglycerols

Lipids are also main components of the human diet. The consumer preference for plant-derived oils is increasing to the detriment of animal fats. Annual plant oil production is increasing worldwide and most of it is used for human consumption as margarines, oils and food ingredients. The triacylglycerols are the most important components of plant seed oils. Interestingly, the physical and chemical properties of an edible oil are related to the chemical structure of the fatty acids esterifying the glycerol . Properties such as melting point, colour, flavour, mouthfeel, spreadability, stability, and effects on human health are determined by the fatty acid composition of the triacylglycerols. Most efforts in developing changes in the lipid composition of plant oils have been directed to change the proportion among the fatty acids of the triacylglycerols.

Common fatty acids in the commercial seed oils are lauric, myristic, palmitic, stearic, oleic, linoleic and linolenic. As is apparent, their differences occur in the length of the carbon skeleton (C12 to C20) as well as in the presence of double bonds are named polyunsaturated fatty acids (PUFA). Different studies on the of consumers towards a reduction of saturated acids in the diet and, accordingly, an increase in unsaturated acids. Epidemiological studies have shown that intake of monounsaturated acids was associated with a low incidence of coronary artery disease, which has been explained by its reduction in the low density lipoproteins (LDL) levels and their oxidation. Therefore the unsaturation of fatty acids have been the target for modification by genetic engineering studies. Although the metabolic pathways leading to the synthesis of these compounds are not simple, some genes have been adequately selected to be modified. Oleic acid, the major monounsaturated acid of the diet, reduces cholesterol and LDL in the serum. Transgenic plants overexpressing the desaturase gene, that encodes for the enzyme catalysing the conversion of the saturated precursor stearic acid C18 into oleic, have been obtained. This has determined that oleic content of soybean has been raised to values up to 80% of the total fatty acids content of the seeds. Although the relationship between PUFA and disease remains contentious, there is a consensus among the health organizations the PUFA should form 8–23% of the total lipid intake in the human diet. However, the production of high PUFA oil plants is not straightforward (Figure. 6.2), and no report on transgenic plants with high PUFA content is known.

TABLE 6.1 : NOMENCLATURE AND REPRESENTATIVE EXAMPLES OF NATURALLY OCCURRING FATTY ACIDS

Systematic nomenclature	m:n $\Delta^{aZbZ...}$ m: carbon atoms n: double bonds superscript: positions of the double bonds a,b,. . . .. caron numbered from the carboxyl end Z: configuration *cis* of the double bond	
Examples:		
Saturated	18:0	Stearic acid
Unsaturated	18:1 Δ^{9Z}	Oleic acid
Polyunsaturated (PUFA)	20:5 $\Delta^{5Z,8Z,11Z,14Z,17Z}$	Eicosapentaenoic acid

Interestingly, there have been cases where saturation of the fatty acids has been the purpose of the plant genetic modification. Saturation of fatty acids determines properties such as melting temperature and viscosity that may be important to a commercial product. Margarine, for example, needs to be easily spreadable within a range of temperatures. In addition, saturation may also be beneficial for oil stability since it is known that unsaturated acids are more readily oxidized, resulting in an increased tendency to rancidity and off odours. Finally, vegetable oils used for frying require partial saturation by hydrogenation in order to give adequate characteristics of stability and melting temperature to these oils. The chemical hydrogenation has been proven to induce also a change in the configuration of the double bonds of the fatty acids, from the naturally occurring *cis* to *trans*. The presence of the *trans* unsaturated fatty acid has been correlated to a risk of coronary heart disease. Therefore, plants with a high content of natural oil and with a high level of saturation have been engineered. High stearate content of the oil in a *Brassica* plant has been achieved by two methods. One method transformed this plant species with the antisense construct of the gene encoding the stearoyl-ACP desaturase, the enzyme that catalyses the transformation of stearic acid into oleic acid (Figure. 6.2, Δ^{9}DS). The silencing of the endogenous gene produced the accumulation of stearic acid up to 40% of the total fatty acids content. The second method transformed the *Brassica* plant with a gene encoding the stearoyl-ACP thiosterase specific for the synthesis of stearic acid (Figure. 6.2). Using this approach, the transgenic plants yielded up to 68% of this fatty acid.

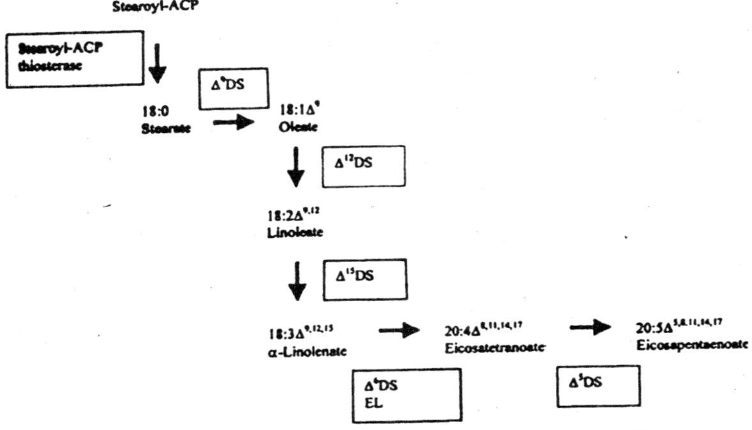

Figure. 6.2: ***Interconversions of the fatty acids indicated in Table 6.1 as examples, Enzymes: Δ ^{n}DS: n-desaturase: EL: elongase.***

Vitamins and Diet Enrichment Compounds

Vitamins are essential compounds for humans and other vertebrates and they must be obtained from the diet. In addition, some vitamins are used as functional additives in food products. Ascorbic acid (vitamin C) is used to prevent oxidation in apples, peaches, apricots, potatoes, peanut butter, potato chips, beer, fat and oils. The carotenoids (vitamin A precursors) are used as colorants in margarine, cheese, ice cream, pasta, juices and beverages. It is generally believed that people in Western countries have adequate vitamin intake. There are, however, susceptible groups within the general population who may have inadequate vitamin intake. Such groups include dieters, people on medication, pregnant women, alcoholics, adolescents, and people with diabetes and other chronic ailments. In developing countries very little information exists concerning their nutritional status.

The edible part of rice grains, the endosperm, lacks and essential nutrient as vitamin A. A diet mostly based on rice consumption may eventually cause vitamin A deficiency. It is estimated that improved vitamin A nutrition could prevent worldwide 1–2 million deaths annually among children. A very promising achievement has been the introduction of genes into rice that enabled the biosynthesis in the endosperm of β-carotene, the precursor of vitamin A. β-carotene is synthesized from the precursor geranylgeranyl bisphophate which is converted to the colourless phytoene by the enzyme phytoene synthase. The phytoene undergoes four desaturations to form lycopene, which is red and gives colour to ripened tomato fruits. Further cyclisation of lycopene results in the formation of β-carotene. Immature rice endosperm is capable of synthesizing geranylgeranyl biphosphate which can be used to produce phytoene by expression of a phytoene synthase gene. The introduction of three genes in rice via *Agrobacterium* allowed the expression of the entire β-carotene pathway into the endosperm. These genes were a phytoene desaturase from *Erwinia uredovora*. The grain of the transgenic rice had a yellow-golden colour and by itself contained sufficient β-carotene for human vitamin A requirements. In rapeseed, transformation with a phytoene synthase gene also increased the level of vitamin A precursor. In tomato the transformation with a bacterial phytoene desaturase increased up to twofold the β-carotene content in fruits.

Another lipid-soluble vitamin whose function is linked to an antioxidant role is vitamin E (α-tocopherol). Daily intake of this vitamin in excess of a recommended minimum is associated with decreased incidence of several diseases. Plant oils are the main source of dietary vitamin E and they generally have a high content of the vitamin E precursor γ-tocopherol. Overexpression of γ-tocopherol methyl transferase greatly increased the seed level of α-tocopherol in the model plant *Arabidopsis thaliana*. This process seems ready to be eventually applied to some commercial crops in the future.

Flavonols are another group of secondary metabolites whose inclusion in the human diet may give protection against cardiovascular diseases. The biosynthetic pathway leading to the

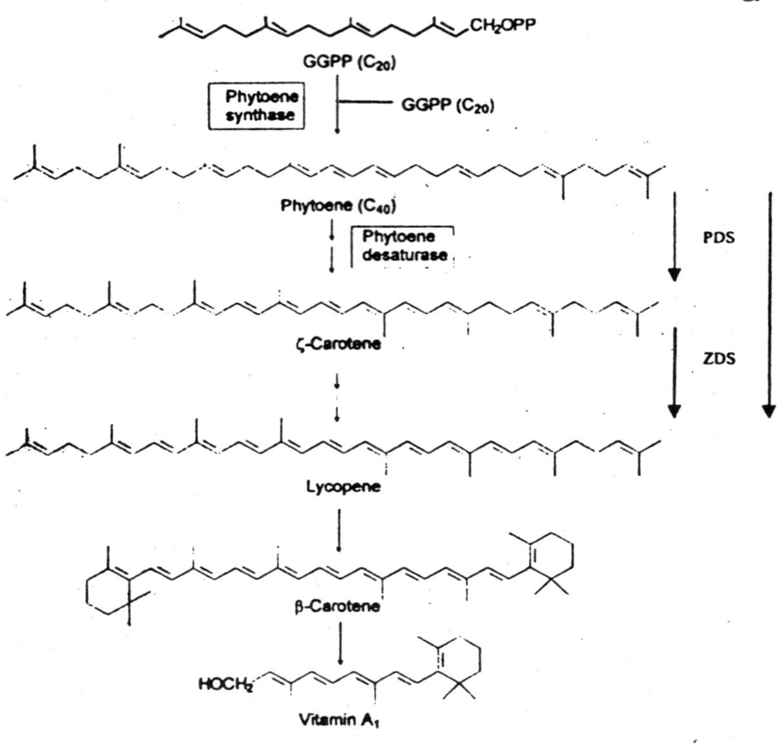

Figure. 6.3: ***Biosynthetic pathway of vitamin A from geranylgeranyl pyrophosphate (GGPP). Vertical arrows indicate steps catalysed by phytoene desaturase (PDS), carotene desaturase (ZDS), and*** **Erwinia uredovora** ***phytoene desaturase (CRTI).***

synthesis of these compounds has been known for a long time. However, recent information regarding the pathway has allowed the design of specific strategies to increase the content of selected bioactive compounds. Thus, the transformation of tomato with a gene from Petunia encoding a chalcone isomerase has produced tomato fruits with a 78-fold increase in the content of flavonols in the peel. What is more important, 65% of the flanonols were retained in the paste obtained after processing the transgenic fruits.

A high risk of iron deficiency has been reported when vegetables are the major components in the diet. Although some plants are rich in this element, its availability is limited by the fact that the same plants contain oxalic acid and phytate-like substances that may complex this element. Some studies have shown that oral administration of ferritin, a protein used by plants and animals to store iron, can provide the iron needed to treat anaemia in rats. With this information, rice has been transformed with a soybean gene encoding ferritin, under the control of a seed-specific promoter. Transgenic rice plants accumulated ferritin in the endosperm tissue and up to three-fold levels of iron in comparison to normal seeds. Interesting, plants overexpressing ferritin have been reported to be tolerant to oxidative damage and pathogens. This seems to be an additional agronomic trait for these transgenic plants.

MODIFICATION OF FRUIT COLOUR AND SWEETNESS

It is becoming widely accepted that plant biotechnology is entering a second phase of development that looks for differentiated crop development. This phase is characterized for being more focused on output traits than input traits. Some of these output traits have been described in the previous section, but still there are others that are related to fruit quality for human consumption. In general, market trends predict growth in the consumption of fruit if quality can be improved. These traits remain a challenge for plant biotechnology developers. However, there are two aspects that are present in the efforts of various research laboratories. They focus on fruit colour and sweetness, and they may be on the market in the near future.

Carotenoids are a group of natural pigments that determine the colour of many flowers, fruits and vegetables. In fact, the golden-yellow colour of high β-carotene transgenic rice has previously been indicated to the a side effect of rice plant transformation. In some cases, the goal has been specifically to increase the colour of the transgenic product. This trait is particularly important in fruits that are going to be processed to produce jams, marmalades, pastes, and even wine. Recently, the carotenoid biosynthesis pathway has been modified in tobacco plants using the *CrtO* gene encoding the β-carotene ketolase, from the alga *Haematococcus pluvialis*.The transgenic plants accumulated several ketocarotenoids that changed the colour of the nectary from yellow to red. The researchers claim that plant transformation with this gene may be used in the future to change the colour of fruit.

Flavonoids are a major class of secondary plant products well known for the colouration they provide in blue, red and purple flowers, fruits and leaves. Flavonoids are derived from simple metabolites as phenylalanine and acetyl-CoA in a highly branched pathway that leads to flavonols, flavanones, isoflavonoids and anthocyanins. Each type may undergo further modifications that result in a great diversity of colours. It is known that this complex pathway is regulated at the level of transcription of structural genes thus giving a main role to transcriptional factors, i.e. proteins that control the expression of some other genes at transcriptional level. This has been applied to the model plant *Arabidopsis* and tobacco, which were transformed with the maize *R* and *C1* genes. These are regulators that produced an activation of genes encoding for enzymes of the anthocyanine production pathway in the transgenic plants. In general, changes at the level of intermediate enzymes of the flavonoids pathway may change the final balance of these coloured compounds and eventually the colour of a given plant organ, for example, the production of yellow flowers in acyanic lines of Petunia after its transformation with a chalcone reductase gene from *Medicago sativa*. Only redirection of the flavonoid biosynthesis pathway could explain the reported effect since chalcone reductase activity is not naturally present in Petunia as it is not the product of the reaction, which is further transformed in a coloured compounds. This illustrates the complex equilibrium of the complete pathway and the difficulty of predictable effects after plant transformation with heterologous genes.

A critical component of fruit palatability is sweetness. There are two goals associated with this food property to be achieved. The first is that many fruits are not sweet enough to make them appetising to consumers. The second is that the most common natural sweetener is sucrose, whose caloric content has caused a change in consumer preference for other sweeteners. Both problems have been addressed by plant biotechnologists.

It was known that a protein from the plant *Dioscorephyllum cumminsii*, named monellin, was about 10^5 sweeter than sucrose on a molar basis. This property is common among some plant proteins of the thaumatin-like class. Tomato plants overexpressing the monellin gene under the control of a fruit specific promoter (E8) have been produced. Although no further reports on fruit properties have been reported, the experience indicates a possible way to increase fruit sweetness, at the time that protein content is enhanced.

Other researchers have also looked for substitutes for the natural sweetener sucrose. They identified the low molecular weight fructans, polymers of fructose, as an adequate replacement. These compounds resemble sucrose in their organoleptic properties, but are indigestible by humans. to obtain high fructan plants, the gene encoding 1-sucrose:sucrose fructosyl transferase from *Helianthus tuberosus* was introduced into sugar beet. Most of the stored sucrose in the root was converted to low molecular weight fructans. The experience is ready to be extended to other commercial crops whose sugar content make them unattractive to consumers.

MODIFICATION OF FOOD-PROCESSING

Wheat is a cereal commonly used to make bread and a number of end-products such as cakes, cookies or pastry. The adequacy of the wheat grains used for each purpose is related to two main properties of the grain that have been consequently targeted as susceptible to being modified by genetic engineering. One is the gluten protein composition that may eventually determine the viscoelasticity (elasticity and extensibility) of the dough. The other is the grain.

What is known as gluten is in fact a complex mixture of up to 50 proteins most of them classified as prolamins. Among these, a group is assembled in high molecular weight (HMW) polymers whose subunit composition determines some properties of the dough, including its elasticity. The entrapment of CO_2 in the formed network favours a porous structure that determines the viscoelasticity of the dough. Since wheat is hexaploid, up to six possible genes for the HMW subunits may be expected. However, it has been demonstrated that the modification of one or two genes may be sufficient to change the properties of the dough prepared from the grain. Thus, wheat transformed with one or two HMW subunits produced grains that caused a stepwise increase in dough elasticity.

Recent studies have demonstrated that two wheat proteins, named puroindolines A and B, acting together in a 1:1 ratio, control to some extent wheat grain softness. In cereals with a hard texture, such as corn and rice, these two proteins are not present in the grain. To genetically modify the texture of these cereals, transgenic plants expressing the wheat puroindoline genes were obtained. The result was that rice grain softness has been increased significantly.

Potato constitutes a basic food in the diet of many Western countries where it is commonly stored in the cold to prevent sprouting. Under this condition part of the starch of the tuber is converted into hexoses. The increased hexoses induce the sweetening of the tuber and have an adverse effect on the quality of the processed chips. The reason is that an excess of hexoses react with the amino acids during frying causing an undesirable browning of the chips. Hexoses increase in the cold stored tubers results from an imbalance between the rate of their production from starch and the rate of their degradation in the glycolytic pathway. The accumulated hexoses may be metabolically converted into sucrose that is eventually split into glucose and fructose by the action of the enzyme invertase. Several strategies have been attempted with variable success to keep the hexose level low in the cold stored tubers. The latest is based on a gene encoding a protein that inhibits the invertase enzyme. Thus, in the presence of this protein, hexoses would not be produced from the hydrolysis of sucrose. Transgenic potato plants overexpressing this invertase inhibitor gene reduced by 70% the hexose accumulation after the cold treatment and, as expected, browning of the chips was also prevented. Most importantly, this occurred without any change in the starch quality or quantity.

Around 30% of the starch produced in plants is used for direct human and animal consumption. The ratio between the two main components of this polymer, linear amylose and branched amylopectin, determines its applicability. The reason is that the amylose:amylopectin ratio influences the physicochemical properties of the starch. This ratio fluctuates among different crops, between 20–30% for amylose to 70%–80% for amylopectin. There has been recently a successful report on the variation of this ratio in a transgenic potato plant. The species was transformed with the antisense of two genes encoding two isoforms of a starch branching enzyme. The result was that amylopectin was practically absent as a component of the starch.

MOLECULAR FARMING

The production of plant-derived biopharmaceutical products is sometimes named as molecular farming. The word biopharmaceutical is applied to a naturally occurring or modified polypeptide, protein, DNA or RNA product that is to be used for therapeutic, prophylactic or *in vivo* diagnostic use in humans. The main categories of biopharmaceutical products are recombinant proteins, therapeutic monoclonal antibodies, polyclonal antibodies, non-recombinant proteins and antisense oligonucleotides. Between 1995 and 1999, 15 recombinant proteins, six monoclonal antibodies, two polyclonal antibodies, two non-recombinant proteins and one antisense oligonucleotide were approved by the FDA. By mid-2000 there were between 80–90 biopharmaceuticals in general medical use, and around 500 more were unrgoing clinical trials. Major targets of these compounds include cancer, cardiovascular diseases, and infectious diseases. Most of these products have been produced in cultured mammalian cells, bacteria and fungi. Now, the use of plants as alternative production systems is being evaluated since plants are potentially a cheap source of recombinant products.

One possible disadvantage of using plants as bioreactors for biopharmaceuticals is that post-translational modification of synthesized proteins may differ from mammals. However, these modifications are few compared with the differences between mammals and microorganisms that have been commonly used as a source of biopharmaceuticals. There is also the risk of impurities in the plant-derived biopharmaceuticals that may include secondary plant metabolites, pesticides and herbicides. Such impurities could have a direct toxic effect, could affect product stability, or could even have immunogenicity leading to allergic reactions. However, biopharmaceuticals derived from transgenic plants could be safer than those derived from human cells that could be contaminated by human pathogens.

TABLE 6.2 SOME EXAMPLES OF PROTEINS WHOSE PRODUCTION IN VARIOUS PLANT SPECIES HAVE BEEN REPORTED

	Potential application	**Plant**	**Protein**
Vaccines			
	Hepatitis B	Tobacco	Hepatitis B surface antigen
	cholera	Potato	V. cholerae toxin Ctoxa and Ctox B subunits
	HIV	Tobacco	HIV epitope (gp120)
	Malaria	Tobacco	Malarial B-cell epitope
Antibodies (single chain Fv fragments)			
	Production of protein in tubers	Potato	Phytochrome binding scFv
	Treatment of non Hodgkin's lymphoma	Tobacco lymphoma	scFv of IgG mouse B-celi
	Production of tumour associated marker antigen	Cereals	scFvagainst carcinoembryogenic antigen
Biopharmaceuticals			
	Anticoagulant	Tobacco	Human proteinC
	Anaemia	Tobacco	Human erythopoietin
	Provitamin A deficiency	Rice	Daffodil phytoene synthase
	Hypertension	Tobacco	Angiotensin-1-converting enzyme

Transgenic plants have already been developed to produce proteins such as enkephalines, α-interferon, human serum albumin, glucocerobrosidase and granulocyte-macrophage colony-stimulating factor. These last two are among the most expensive drugs before their production in plants. Rice plants have been engineered to produce α-1-antitrypsin and the product is presently under trial. In one case the human somatotropin has been reported to be produced in the chloroplast of a non-food crop such as tobacco claiming the additional advantage of biological containment in the field cultivation of this plant.

Among possible proteins to be produced in transgenic plants there have been great efforts to produce antibodies whose therapeutic potential has been largely recognized. Functional antibodies have already been produced in plants and sometimes have been named as plantibodies. Plants have been successfully used to generate complex secretory antibodies that would be of particular benefit for topical immunotherapy in mucoses. This in the case with the production of humanised monoclonal antibodies for immunoprotection against genital herpes. Moreover, the use of edible plant parts as a source of abibodies opens the door to the possibility of treatment of the mucoses of the gastrointestinal tract.

The development of plants expressing vaccine antigens is another relatively new potential application of plant biotechnology. Vaccines consisting of macro-molecules with a protective immune response has a limited use in developing countries, mainly owing to their high cost and low stability. The low requirements of growing plants make their production much cheaper. The expression of vaccines in plant tissues also eliminates the risk of contamination with animal pathogens, make them more stable, and may allow oral delivery if expressed in edible parts of the plants. The first clinical trial with a plant-derived vaccine in 1997 demonstrated the induction of a mucose immune response. Potatoes genetically modified to produce a cholera toxin B subunit has been shown to induce antibody production in humans after oral administration. Also, in preclinical animal trials, mice fed with transgenic potato expressing hepatitis B surface antigen results in a primary immune response. More recently, three plant synthesized antigens of cholera, rotavirus and enterotoxigenic *E. coli* were expressed in potato and showed a strong immune response in potato-fed mice. Since it is well documented that delivery of plant-derived vaccine to a mucosal site induces both local and systemic immune responses, the list of plant-derived vaccinogens continues to grow, and includes viral, bacterial, enteric and non-enteric antigens.

FUTURE TRENDS

Experts in the field indicate that it is most likely that Biotechnology will play a significant role in the 21st century. At present, most of the commercial applications of transgenic plants have been for crops with improved agronomic benefits such as resistance to pests or herbicides. This trend will continue in developing countries since there is still the possibility of increasing the yield ceiling. However, it is likely that the focus for development in the coming years will be in multiple gene introductions to increase output traits such as increased nutritional value, vitamin content, or improved flavour components. Specifically, in the area covered by this chapter, the main trend pointed out by experts is in the production of nutraceuticals. This term was coined in 1979 for 'foods, or parts of foods, that provide medical or health benefits, including the prevention and treatment of disease'. The term includes some other definitions such as dietary supplements, functional foods and medical foods. It was estimated that by 1998, 47% of the population of Japan consumed nutraceuticals. Although there is no agreement among experts on the size of the world market for these products, it seems that the trend to a higher consumption is expected. Nutraceuticals include a broad range of products from high-tech foods, such as engineered canola with increased levels of antioxidants, to low-tech candy bars supplemented with vitamins. In plants, the number of possibilities for increased demand from consumers can only be estimated at present. The abundance of basic information, the improvement of plant transformation and protein delivery methods, the development of computational biology, bioinformatics, and the trend of individualized medicine will mark the direction of this area in the future.

7

Nutritional Enhancement of Plant Foods

The question of why it is necessary to improve the nutritional value of plant foods is one that at first hand might seem difficult to justify. What evidence is there that this is a problem? In the developed world there are no overt signs of malnutrition even amongst strict vegans. The reasons for this are that many processed plant foods are fortified with essential nutrients. Fortification is utilized to replace nutrients lost in the heat processing of foods and through oxidation. Few vegetarians are dependent on a single plant source to provide their basic nutritional needs. In addition, vegetarians frequently consume vitamins as supplements and the growth in this industry has been rapid. The fact that people are resorting to the consumption of vitamins as supplements is a reflection of their belief that more of a good thing will result in an improvement in their health. This is a very dubious argument. Nonetheless, it is important to recognize that the recommended intakes of nutrients, that have been determined by expert groups of nutritionists, are based on the evidence that a specific intake level for a nutrient is required to ensure healthy growth and development. They do not reflect the growing body of evidence that suggests different, and often higher, intakes of these same nutrients are required to optimise health and lead to an active life through the prevention of chronic degenerative diseases associated with ageing. The critical issue is to determine what intakes are required to optimise health rather than to compromise it.

THE NUTRITIONAL IMPORTANCE OF PLANTS

Plants are the staple food for the vast majority of the world's population. It is known that many staple plant foods are deficient in essential nutrients and, consequently, malnutrition is widespread. It has been estimated that over 100 million children worldwide are vitamin A deficient and improving the vitamin A content of their food could prevent as many as two million deaths annually in young children. This is apart from the deficiencies in iodine intake, resulting in goitre, and in iron-deficient anaemia which are estimated to affect millions in the developing world. There is also an important need to improve the amino acid content of legume proteins that are deficient in essential sulphur amino acids. Nutritional deficiencies can lead to a reduction in immune responsiveness, rather than a specific attributable disorder, making it difficult clearly to establish how many people are suffering from malnutrition.

In the developed world all public health authorities are urging consumers to consume more plant-based foods as part of a healthy diet. There is a significant body of evidence to suggest that the traditional Mediterranean diet, rich in plant foods, reduces the risk of many age-related diseases. Epidemiological studies show a strong and consistent inverse relationship between fruit and vegetable intake and the risk of cardiovascular diseases and some cancers. An explosion of interest in trying to define what are the factors in fruit and vegetables which might be responsible for these observations has not yet led to a clear set of explanations although many theories abound.

Plants contain 17 mineral nutrients, 13 vitamins and numerous phytochemicals that have been shown to have potentially beneficial effects on health especially against the initiation or progression of degenerative diseases. Almost all human nutrients can be obtained from plant foods, the exceptions are vitamins B_{12} and D. However, the adequacy of a plant diet in delivering a health benefit from a specific component will depend on the amount ingested, and its bioavailability. Many beneficial plant compounds that are associated with the plant cell wall are not easily bioavailable. Any way in which overall levels can be increased will help overcome this difficulty.

STRATEGIES FOR NUTRITIONAL ENHANCEMENT

- the application of traditional breeding methods to select for varieties with an increased level of the bioactive compound
- a reduction in the content of antinutritional factors
- the use of genetic manipulation to introduce new traits in plants
- improvements in handling, storage and food processing technologies.

Each of these approaches has a role to play but genetic manipulation provides a mechanism for the improvement of nutritional quality that overcomes the problem of the absence of a specific biochemical pathway in a staple crop.

Application of 'Traditional' Breeding Methods

Plant varieties have not been selected to date on the basis of nutritional qualities but there are wide natural variations that can be found in the gene pool of crop plants. Examples of where significant variations in the nutrient content of genotypes have been documented include a:

- 2-fold variation in calcium concentration in beans
- 4-fold variation in β-carotene concentrations in broccoli
- 4-fold variation in folates in beetroot
- 2–3 fold variation in iron and zinc levels in maize

In the case of the pro-vitamin A carotenoids, plants provide highly variable amounts depending on their colour. Varieties of sweet potato may contain levels varying from 0.13mg to 11.3mg g^{-1} dry weight of β-carotene. Similar variations in levels can be found in carrots and cassava. In the case of the tomato, genes have been identified that are associated with high and low lycopene content. Incorporation of genes that increase lycopene content and/ or elimination of genes that decrease the lycopene content, can be achieved by pedigree selection and backcross programmes. Such techniques have produced hybrids with a three- or four-fold content of lycopene in tomato fruits.

Reduction in Antinutritional Factors

The interest in reducing antinutritional factors in plants has been predominantly focused around improving the nutritional value of feedstuffs. Phytates are present in many plant seeds and limit phosphorous uptake as well as other elements. The potential for introducing a phytase gene into feedstuffs has been explored. However, there are other strategies that seem to be of greater overall value in human nutrition. Thioredoxin is thought to be an activator of the germination process in seeds. It is able to activate proteins to degradation by proteolysis and results in improved digestibility. It also has the potential advantage of being able to reduce allergenicity, presumably because of its capacity to break disulphide bonds by the action of the reduced thiol groups in the molecule and ensure the tertiary structure of the protein is accessible to degradation by proteases. The insertion of the wheat thioredoxin gene into barley has produced a transgenic plant where thioredoxin accounts for 7% of the total protein content in the barley and is a good source of sulphur amino acids.

The Application of Genetic Manipulation

Genetic engineering is being applied to enhance levels of functional compounds in food crops. Indeed for some purposes it will be the only approach feasible especially where there are widespread deficiency diseases and the population is dependent on staple crops which are not source of the nutrient required. There are many examples where technology has been applied with success although there are no products which have yet reached the marketing stage where nutritional benefits have been the main focus.

Potential strategies for the enhancement of specific metabolites could target on:

1. over expression of enzymes that control the final steps in the biosynthesis of a metabolise
2. over expression of rate-limiting enzymes
3. silencing of genes whose expression causes the metabolise to be degraded
4. increased expression of genes that are not subject to metabolic feedback control
5. increasing the number of plastids in a plant
6. increasing metabolic flux into the pathway of interest
7. expression in storage organs using site-specific promoters.

The strategy that has had the greatest success at present is the first one, especially in conjunction with the last strategy. In practice if a substantial increase in the concentration of a metabolite is required, the use of specific promotes directing the synthesis to a particular organelle normally used for storage purposes, or where the plant normally synthesizes the metabolite, is essential. Failure to use these could cause toxicity in the plant by interfering with the production or function of other essential metabolites. However, this strategy presupposes the metabolite of interest is the final one in a particular pathway.

Few strategies have yet been applied where multiple gene insertions are necessary to produce the metabolite, although these are progressing rapidly, and none where plastid numbers have been increased. However, the accumulation of sequence data of both chromosomal DNA and expressed sequence tags of plants and other species is providing rapid advances in knowledge of the genetic make up and functions of several plants and it is expected that these other possibilities will soon be feasible.

NUTRITIONAL ENHANCEMENT

For the Developed World

Although it is known that the distribution and processing of food can lead to a significant loss in nutritional quality, there are few instances where present evidence suggests there is a need to change current practices. There is very little evidence for nutritional deficiencies. In those cases where public health authorities have thought there is a potential problem food supplementation with nutrients is a commonly adopted policy. The use of nutritional supplements is widespread. Whilst the focus of current interest is on the need to consider nutrients and other phytochemicals as protective against the development of disease in later life, the levels of intake that may be necessary to optimise protection are far from resolved at the present time.

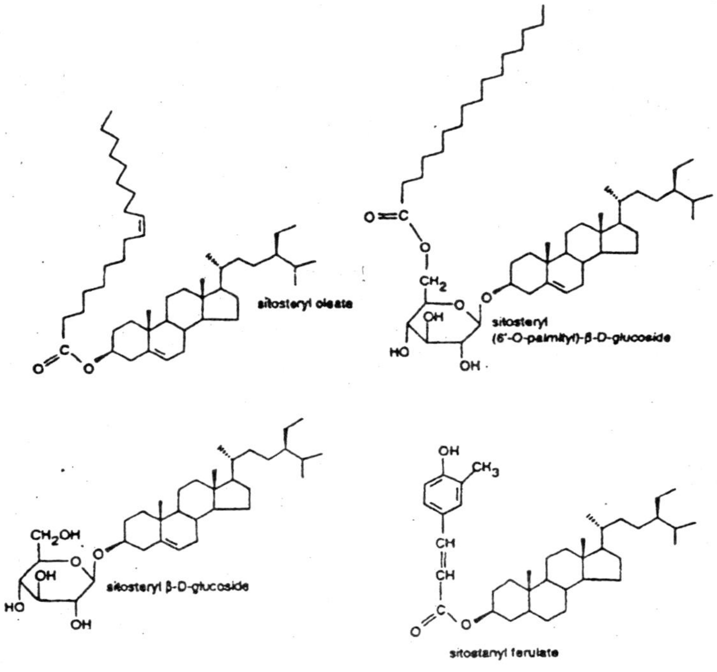

Figure. 7.1: ***Structure of plant sterol and stanol esters.***

The only plant-derived food product on the market where nutritional health benefits are claimed (as opposed to implied) is the enrichment of margarines with plant sterol and stanol esters for the reduction in plasma cholesterol levels (Figure. 7.1). These products do not require the development of specifically bred plants since it is possible to extract stanols and sterols from existing plants (albeit in the case of the stanols from the bark of a tree) for use in their manufacture.

Experiments with plant stanol esters were shown to lower serum cholesterol consistently by about 10–15% and LDL-cholesterol by about 20% in patients with high serum cholesterol levels as well as in normal individuals. Similar effects have been seen with plant sterol esters but at least 1g/day of plant sterols need to be consumed. Consequently they require extraction and addition to foods.

Plant sterols can be in the free form or predominantly esterified with long chain fatty acids or with phenolic acids as in rice-bran oil (ferulate) and shea butter (cinnamates). Sterol esters are better absorbed than the free sterols and most sterol esters are hydrolysed to the free sterols in the intestine.

As campesterol esters are better absorbed than sitosterol esters, serum levels of campesterol could rise to those levels that are found in the very few people who suffer toxic symptoms from phytosterolemia. Thus there may a benefit in increasing the sitosterol to campesterol ratio in plants.

The ideal situation would be for sufficient sterols to be present in our diets to ensure that plasma cholesterol levels are kept reasonably low without the need to buy a specific functional food, and that they would be in a fat soluble form for effective uptake. The evidence favours in increasing order of preference the use of:

1. plant sterol esters with low campesterol contents
2. sterol esters from tall oil (derived from pine wood) which have a higher stanol content than edible oils
3. plant stanol esters.

A vegetable oil rich in plant stanols, especially in sitostanol esterified with polyunsaturated fatty acids, would also have the benefit of being less susceptible to oxidation at frying temperatures than the sterols. The potential health benefits of this class of bioactive compounds are unlikely to be met by the use of classical plant breeding methods but genetic engineering could make these targets feasible.

For the Developing World

The world-wide deficiency of vitamin A is being tackled both through conventional plant breeding and by genetic manipulation. However, the use of conventional plant breeding to deliver adequate intakes is dependent on availability of carotenoid-rich staple foods. Often these are available for very restricted times of the year in some societies. In those countries where rice is a dietary staple the problem is particularly severe and the deficiency is likely to be corrected only by the introduction of rice that has been genetically manipulated to produce β-carotene. However, yellow rice is produced and this may give rise to problems of acceptability to consumers used to white rice.

Manipulation of the carotenoid pathway in rice.

The nature of the challenges faced in manipulating plant secondary metabolites is well illustrated through the attempts that have been made to produce carotenoids in rice plants. A simplified version of the pathways leading to the synthesis of the carotenoids principally found in food plants is shown in.

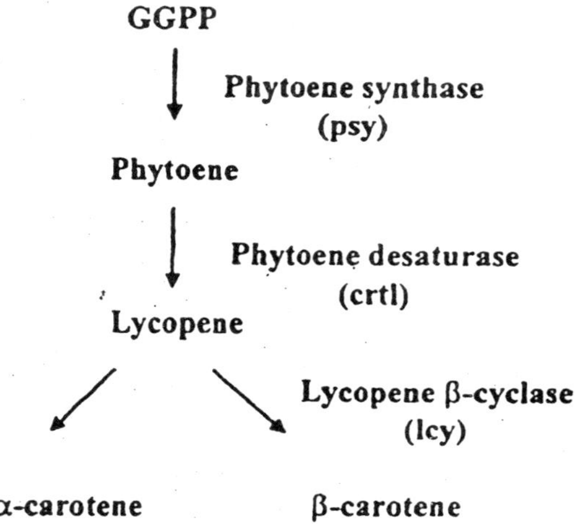

Figure. 7.2: ***The carotenoid biosynthetic pathways (simplified).***

Immature rice endosperm is capable of synthesizing the early pathway intermediate geranylgeranyl diphosphate (GGDP). Four plant genes corresponding to the enzymes phytoene synthase (psy) (1), phytoene desaturase (2), zeta carotene desaturase (3) and lycopene cyclase (crt) (4) are required. Enzyme (1) was obtained from the daffodil (Narcissus pseudonarcissus), (2) from a bacterium Erwinia uredovora – which is capable of achieving steps (2) and (3) from the single enzyme, and (4) from the daffodil.

The genes need to be expressed in a tissue-specific manner through the insertion of specific promoters. This has been achieved in rice through the use of the daffodil psy gene. In rice the daffodil psy cDNA insertion is under the control of an endosperm-specific promoter. The choice of promoter will very much affect the timing and tissue-specific expression of a gene.

Surprisingly, seeds that expressed psy and crt did not accumulate lycopene. Instead they contained β-carotene and other xanthophylls. Thus is would seem that the enzymes required to make these metabolites are either normally expressed in rice endosperm or are induced if lycopene is formed. The maximum level of carotenoids in the endosperm of plants that were heterozygous for the transgenes was 1.6 mg kg^{-1} which is likely to help to meet the nutritional needs of people consuming rice as a staple. Interestingly, good progress is being made in adding a gene coding for ferritin – the iron storage protein found in mammals and plants – to rice. It is likely that this is shown to be bioavailable.

The controversy over the use of advanced technologies for producing sustainable food in the developing world has been addressed by the developers of modified rice. They have in effect waived all intellectual property rights for exploitation of the technologies in the developing world, and are actively involved in assisting the International Rice Research Institute to breed stable and agronomically successful lines for use in vitamin A-deficient areas.

(BIOAVAILABILITY)

The overall content of a given nutrient in a food is not always a useful indicator of its nutritional value as not all of the nutrient present is absorbed. Nutritionists must concern themselves with understanding the proportion of an available nutrient that is digested, absorbed, and ultimately utilised. In the case of nutrients or phytochemicals, whose beneficial effects are directed towards inhibiting degenerative diseases, it is important to know whether or not the nutrient is reaching the particular target organ and in a form which is active. Otherwise the claims for the health benefits of that chemical would not be justified, especially as it is difficult to demonstrate benefits from long-term human studies.

Diet plays an important role in the uptake of specific nutrients and phytochemicals. Those that are lipophilic are absorbed much more readily from a lipid-rich diet. Frying tomatoes in oil dramatically improves the uptake of lycopene compared with the consumption of fresh tomatoes. Raw carrots, which have high levels of pro-vitamin A carotenoids, are poorer sources of β-carotene than gently cooked carrot. The bioavailability of certain trace elements is increased on cooking or processing; for example, the bioavailability of iron is increased in canned spinach.

The chemical form of the phytochemical present in food is very important in determining uptake through the gastro-intestinal tract. Quercitin-β-glucoside is more easily absorbed than the aglycone quercitin. Isorhamnetin-β-glucoside, which is chemically similar to quercitin, differing only by a single methoxyl group, is more readily absorbed. Flavonoid rutinosides (rhamnosyl 1–6 glucosides) are less easily absorbed. Thus, while some phenols might be better antioxidants than others when tested in *in vitro* systems, this is so little significance in terms of health relevance. What matters is whether the compounds are easily absorbed, are not quickly degraded in tissues, and are able to reach the target sites. Flavonoids that are not absorbed undergo extensive degradation by gut microorganisms, and may play only a limited role in preventing oxidative damage in the colon.

NUTRITIONAL ENHANCEMENT

The importance of enhancing the levels of a natural protective constituent in plant foods is well illustrated in the case of the folates. There is a good chance that folate status even in affluent countries is not optimal. The most important sources of folates in the diet are liver, products derived from yeast, eggs, green vegetables, legumes and certain fruits. Plant foods (vegetables, fruits and potatoes) are by far the single largest contributor to the overall folate intake of adults. Some 40% of the total folate intake is from fruit and vegetable consumption in these countries even when the average consumption is not very high.

Folates have the effect of reducing the levels of plasma homocysteine which is a sensitive biomarker of folate status. A variety of studies have suggested that increased plasma homocysteine levels are a risk factor for cardiovascular disease and stroke. Human studies have shown that if individuals consume a supplement of 100 μg/day of folic acid their plasma homocysteine is reduced to a level of about 7.0 μM/1. Increasing the intake of folic acid beyond that level has no further effect. However, the bulk of the population have homocysteine levels in excess of 7.0 μM/1.

Folic acid is not the natural form of folate that is found in plants where natural folate consists of ten different polyglutamate complexes. Folic acid is however the form of folate that is used in the fortification of food as it is more stable. It is also found to be more bioavailable. Natural folates show only 50% or less of the bioavailability of folic acid.

There is good evidence that to achieve the ideal level of plasma homocysteine dietary levels of folate (as supposed to folic acid) would have to increase from the current average of 200 μg/day to 600 μg/day. This increased intake is also likely to have an important impact on the reduction of Neural Tube Defects (NTDs). Women with a low folate status (about 150 μg/1 red cell folate) have a 0.7% risk of NTDs in their offspring, whereas supplementation with folic acid at does of between 100–200 μg/day, resulted in red blood cell folate levels that have been associated with an optimal reduction in NTD incidence. Since average intakes of natural folates are about 100 μg/day from the diet it would require at least 500 μg of natural folate to be consumed (preferably 600 μg/day) to ensure that the incidence of NTD in the population was kept to a minimum.

The fact that supplemental folic acid can achieve these same effects whilst being more stable and bioavailable would imply that there was little purpose in supplementing natural levels of folate. This ignores the intrinsic difference between the cellular metabolism of synthetic pro-vitamin folic acid compared to the natural folates. The mucosa converts all of the natural forms of folate into 5-methylenetetrahydrofolate monoglutamate. This reaction also occurs when folic acid is consumed but the difference is that for folic acid the process can be saturated at around 300 μg. Intakes in excess of this cause un-metabolised folic acid to enter the circulation. The control of how much natural folate is taken up and retained by cells is regulated by the enzyme methoinine synthase which acts on 5-methyltetrahydrofolate to conjugate it into a polyglutamate which is then retained in the cell. Conversely folic acid does not pass through the methionine synthase pathway and can be conjugated directly, retained and metabolised.

The ability of folic acid to bypass an important regulatory step is that excess cellular levels cause DNA biosynthesis in vitamin B_{12}-deficient cells in cases of pernicious anaemia via the DNA cycle. This causes a haematological response with the risk that the anaemic state is masked and the associated neuropathy is not avoided. Natural folates on the other hand will be poorly metabolised by vitamin B_{12}-deficient cells enabling the anaemia to be detected at an earlier stage. Another concern that has been raised against increasing population levels of folate is that the increased capacity to cause DNA biosynthesis could promote tumour growth. This would be expected to be more of a problem with folic acid than natural folates because of folic acid's less controlled uptake into cells.

CONSTRAINTS ON INNOVATION

The potential fully to exploit GM technologies is severely limited by constraints on the use of the technology itself, as well as in satisfying the legislation that exists on the pre-market approval of foods that have been produced by the technology, or are in some way novel. These constraints are so severe in Europe that in very few cases will any producer see a return on their investment if nutritional improvement is their goal. This market is also affected by the widespread addition of specific nutrients as additives to certain processed foods. Enhancement of any component considered to be beneficial is likely to be of market value only if positive claims can be made. Whenever possible 'conventional' plant breeding will be used.

Genetic Manipulation

There has been a *de facto* European Union moratorium on the approval of GMO products since October 1998. Eighteen products have already been approved under the general EU Directive (90/220/CEE) whilst 14 are pending approval. Five Member States have temporarily banned already approved GM products, which is permitted under the Directive. Two new EU labelling regulations have been drafted but have not been implemented because of a lack of testing methodologies, certifying labels and inspection procedures. The ultimate intention is to ensure that products can be labelled GM free to enable consumers to make an informed choice. It has been argued that products labelled as containing products derived from GM will convey negative messages to consumers. This is likely to be so in the absence of benefits that are clearly seen by consumers. This will occur if plants are used as factories for the production of vaccines and pharmaceutical products. However, it is unclear at present where the benefits will lie in the nutritional field other than for the developing world.

The lack of public confidence in the European food safety system is already causing harm to markets in the US and in developing nations where the technology is already embraced. This is likely to lead to major problems in international trade unless it is resolved.

Safety

No scientific development in food can ignore the very strict regulatory controls that exist before any new or 'novel' product or process can be applied in its production. Food plants produced by 'conventional' plant breeding techniques in general are not subject to any regulatory controls. In some countries voluntary codes of practice have been developed within the plant breeding sector when it was discovered that varieties of potatoes, with good agronomic characteristics, were found to contain high levels of toxic glycoalkaloids.

At the present time, genetically–modified (GM) foods are regulated applying the concept of 'substantial equivalence'. This concept is applied as the basis from which to determine the extent of the requirements for food safety assessment. It a genetically modified food can be characterised as substantially equivalent, it can be assumed to pose no new health risks over its conventional counterpart and can be marketed without the need to undertake extensive toxicological and nutritional studies to determine its safety-in-use.

The principle of substantial equivalence was adopted into the EU Regulation on Novel Foods and Novel Food Ingredients. The Regulation excludes from its controls foods and food ingredients obtained through traditional propagating or breeding practices and which have a history of safe use. GM plants are considered as 'novel' under the terms of the Regulation. However, the detailed safety evaluation provisions of the Regulation do not apply to foods produced by genetic manipulation 'if on the basis of the scientific evidence available they are substantially equivalent to existing foods with regard to their composition, nutritional value, metabolism, intended use, and the level of undesirable substances present'. The Regulation regards food as 'novel' if the characteristics of the food differ from the conventional food regarding the accepted limits of natural variation of such characteristics. It is clear that most nutritionally enhanced plants would be caught under the definition of a 'novel' food.

The principle of substantial equivalence is vague and difficult to define in many cases. Consequently the whole issue of regulation of GM foods is under intensive debate. Meanwhile the EU has applied a *de facto* moratorium on GM plant introductions. The US attitude to regulation has so far been to regard safety as an issue that relates to the characteristic of the food and not to the process (es) that lead to it. Novel food products, of which product produced by GM are included in the definition, are not subject to any specific approval on safety grounds if the constituents of the food are the same, or substantially similar, to substances currently found in other foods.

It is clear that it is never going to be possible to argue that a GM plant is safe any more than to be able to argue that a plant produced by conventional plant breeding is safe. The very concept can be addressed only in the context of a history of safe. The very concept can be addressed only in the context of a history of safe use as a human food. Clearly, the overwhelming evidence supports the view that health benefits arise as a consequence of the regular consumption of a variety of fruits and vegetables, few if any of which have any close compositional relationship to the wild types from which they were bred. Similarly their production, storage and distribution has depended on the use of a wide range of chemical fertilizers and pesticides. These chemicals are extensively rested for safety before approval is given for their marketing and use but this has not removed the widely held view amongst consumers that 'organic products' are better for your health. There is no evidence to support this view and any adverse health effects that there might be as a consequence of the use of pesticides appear to be outweighed by the beneficial effects from the consumption of fruit and vegetables. What determines 'safety' is the overall effect of consumption over a period not the effects of a specific chemical that might be present.

The issue of 'safety' in the context of the ability to market foods which are 'novel' is emotionally charged and without a solid scientific base. Consequently it is unlikely that any industry would want to take on these issues unless they had a product with a potential large market.

8

Stress Tolerance

Crop plants are subject to a range of external factors that adversely affect their growth and development. In this chapter, we will look at the effects of environmental conditions such as temperature, water availability and salinity on crop plants. shows a number of physical factors that may impose on abiotic stress on plants and adversely affect their growth and development. These include a number that can be grouped together as temperature stresses (heat, chilling and freezing), which in turn belong to a larger subgroup that can be categorized as stresses that result in water-deficit. The figure also emphasises the point that most abiotic and biotic stresses directly or indirectly lead to the production of free radicals and reactive oxygen species (ROS), creating oxidative stress.

The impact of abiotic stresses on crop yield compared to biotic stresses (weed, pest and disease effects). One of the first things to notice is the large difference between the average yields of crops and the record yields (an indicator of the maximum yields possible under ideal conditions). It is clear from this data that the major difference between record yield and average yield is accounted for by abiotic stress. Thus, the variation in environmental conditions from one year to the next produces such a variation in yield for wheat that the average yield is only 13% of the maximum.The control of biotic stresses in industrialized farming is such that they tend to reduce the annual yield by a fairly stable proportion, which is generally less than the most adverse abiotic stresses. Improving the tolerance of crops to abiotic stresses could therefore enable them to maintain growth and development during the normal fluctuations of adverse conditions, and consequently buffer crops against the large swings in yield experienced from one year to the next.

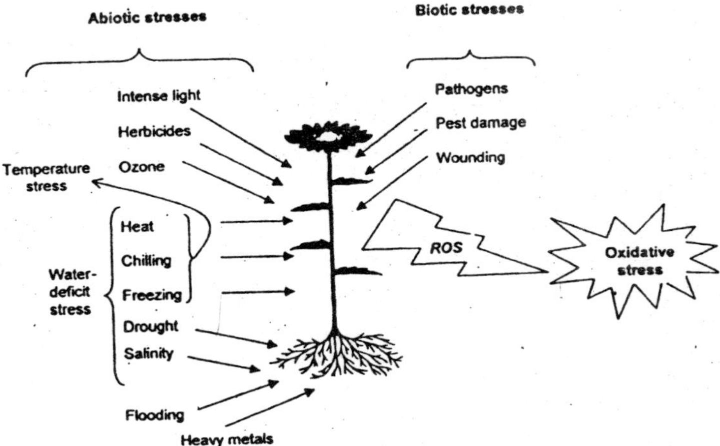

Figure 8.1: ***The different types of external stresses that affect plant growth and development. The figure shows some of the external stresses imparting upon plant growth and development. The stresses are gouped according to common characteristics. Thus, abiotic and biotic stresses are defined, and the range of abiotic stresses resulting in water deficit are shown. The figure emphasises the point that virtually all stresses result in the production of reactive oxygen species (ROS) and this creates oxidative stress.***

In the longer term, the predicted depletion of the ozone layer and climate changes associated with global warming are likely to add to the burden of enviromental stresses on crop plants, and increase the imperative to develop stress-tolerant varieties, Furthermore, there is increasing pressure to extend the area of crop cultivation to environments which are not optimal for the growth of major crops (e.g. desert or high-salt conditions). The development of stress-tolerant plants is therefore a major goal of crop biotechnology, and one that is likely to become increasingly important.

THE NATURE OF ABIOTIC STRESS

When discussing the subject of stress tolerance, it is necessary first to try to define stress in relation of plant physiology. Plants are subject to many types of fluctuations in the physical environment. Many of the strategies used by animals to avoid the effects to these fluctuations are not available to plants, because of the sessile nature of their growth habit. Plants therefore depend largely upon internal mechanisms for tolerating variations in the external environment. Not all such fluctuations present a stress to plants, since they are able to cope with normal variation by virtue of their plasticity (the ability of alter metabolic profiles and developmental trajectories in response to external conditions). Thus, plants are adapted to function in a fluctuating environment, and normal external changes are countered by internal changes without detriment to growth and development. It is only acute or

TABLE 8.1 : AVERAGE AND RECORD YIELDS OF SOME MAJOR CROPS

Crop	Record yield (kg/ha)	Average yield (kg/ha)	Average yield (% of record yield)	Average loss (% of record yield) Biotic	Average loss (% of record yield) Abiotic
Wheat	14500	1880	13.0	5.0	82.1
Barley	11400	2050	18.0	6.7	75.4
Soybean	7390	1610	21.8	9.0	69.3
Corn	19300	4600	23.8	10.1	65.8
Potato	94100	28300	30.1	18.9	54.1
Sugar beet	121000	42600	35.2	14.1	50.7

chronic extremes of environmental condition that lead to environmental stress which has the potential to cause physical damage to the plant. Note also that the 'extremes of environmental conditions' are defined by the normal environment of the particular plant species in question—desert plant do not prosper in the 'mild' climate of the temperate regions.

Given the range of abiotic stresses to which plants are exposed, it might be thought that a wide range of different strategies would be required to engineer particular types of stress tolerance. However, many different stresses cause similar types of damage to plants. This chapter will concentrate on the two major types of damage that result from a variety of different stresses. One of these is the damage that results from water deficit caused by a number of different environmental conditions, including drought, salinity, heat and cold. The other type of damage results from the production of reactive oxygen species (ROS), causing chemical damage to the cellular constituents of plant. Most abiotic and biotic stresses cause the production of reactive oxygen species either directly, or indirectly.

THE NATURE OF WATER-DEFICIT STRESS

Several of the physical changes shown in result in a water deficit, i.e. a situation in which the demand exceeds the supply of water. The supply is determined by the availability of water in the soil to the roots. The demand for water is determined mainly by the plant transpiration rate. Transpiration, the loss of water from the aerial parts of the plant via the leaf stomata, is required to dissipate the energy from solar radiation and the ambient air temperature. Thus, transpiration causes leaves to cool relative to the ambient temperature when the environmental energy load on the plant is high. The rate of transpiration is also affected by the relative humidity and wind speed.

Two parameters are used to describe the water status of plant cells and tissues: water potential and relative water content. Water potential is measured in units of pressure (M pascals or MPa) and is similar in concept to electrical potential—current flows from a compartment of high potential to one of low potential if a connection is made between them. Thus, water is driven through the plant from the soil to the atmosphere (very low potential) and soil (relatively high potential when wet). As water transpires from the leaf, leaf water potential is reduced; but if water is available in the soil, water will flow into the leaf to replenish the loss. However, if the soil water potential starts to reduce (becasue the soil is drying out, for example), a deficit between the supply from the root and demand from the leaf is created. Clearly, a continued water deficit would lead to dehydration and death, but plants can respond to water deficit in a number of ways. One way is to reduce the water potential in the leaf in order to create the necessary gradient to maintain the flow of water from soil.

Two major factors determine water potential: the concentration of solutes that creates the solute- or osmotic potential; and the physical pressure of the cell or tissue boundaries that forms the pressure- or turgor potential. The osmotic potential is generally lower (more negative) than the water potential, and the turgor potential is the difference between them. Leaf turgor is lost (turgor potential = 0) when the leaf water potential drops to a value equal to the osmotic potential. As leaf turgor falls, the stomata close to reduce transpiration. When turgor is lost, the cells collapse and the leaf wilts. These effects have the result of reducing water loss from the leaf, by closure of the stomata, and because wilting or rolling up of leaves reduces their exposure to sunlight.

However, the reduction in stomatal conductance also causes a reduction in photosynthetic assimilation, thus reducing growth. There may also be an increase in leaf temperature to a level that could cause heat damage to the leaf. The maintenance of turgor and transpiration are therefore important in order to maintain the growth of plants under drought stress. Turgor could be sustained either by keeping the leaf water potential high (by maintaining water uptake from the drying soil) or by reducing the osmotic potential. The latter could be achieved by solute accumulation (remember that an increase in solute concentration creates a more negative osmotic potential, i.e. a steeper gradient for the movement of water)—a process called osmotic adjustment.

Different Abiotic Stresses Create a Water Deficit

Drought is, by definition, a water-deficit stress, because the environmental conditions either reduce the soil water potential and/or incerase the leaf water potential due to hot, dry or windy conditions. High-salt conditions can result in water deficit because the soil water potential is decreased (because the osmotic potential of a salt solution is lower than that of water), making it more difficult for roots to extract water from the environment. As stated above, high ambient temperatures cause increased water loss by evaporation, so heat stress creates a water deficit. However, freezing temperatures also cause osmotic stress because the formation of ice crystals in the extracellular space reduces the water potential and results in the efflux of intracellular water. In general, these various causes of water deficit result in the efflux of cellular water, leading to plasmolysis and eventually cell death. Water deficit in a particular problem to plants cells because it inhibits photosynthesis via its effect on the thylakoid membranes. Water deficit is also potentially damaging to all cells because of the increase in the concentration of toxic ions and the loss of the protective hydration 'shell' around vulnerable molecules.

As described above, cells are able to accommodate some fluctuation in the water potential by osmotic adjustment—increasing the solute concentration and hence reducing the osmotic potential. Increase in ions such as Na^+ and Cl^- would be counterproductive, given their effects on vulnerable molecules. Instead, plant cells respond to osmotic stress by producing non-toxic compounds called 'compatible solutes' (or 'osmolytes', or 'osmosprotectants') to reduce the osmotic potential. It has been proposed that one reason for the non-toxicity of these compounds is that they can accumulate to high intracellular concentrations without disrupting the hydration shell around proteins and membranes.

a range of compatible solutes produced by plants. These fall into two broad classes: sugars and sugar alcohols; and zwitterionic compounds (i.e. carrying a positive and negative charge). The former class includes sugar alcohols such as mannitol, sorbitol, pinitol and D-ononitol, and oligosaccharides such as trehalose and fructans. (The production of some of these compounds in transgenic plants). The latter class includes amino acids such as proline, and quaternary ammonium compounds such as glycine betaine. Different plant species tend to produce different compatible solutes. For example, mannitol is produced in large amounts by celery in response to salt stress, whereas spinach accumulate glycine betaine. In fact, spinach cells grown in a high-salt medium accumulates glycine betaine to concentration of 3000 mmol l^{-1} in the cytosol. This maintains the osmotic potential of the cytoplasm relative to the apoplast (the connected extracellular space outside the cell; the symplast is the continuous intracellular compartment without protoplasts, connected by plamodesmata)

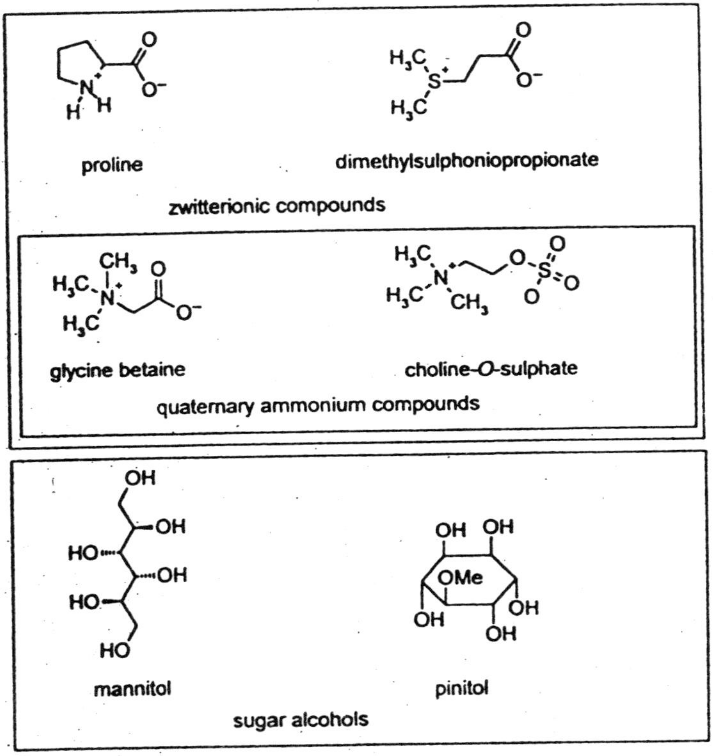

proline dimethylsulphoniopropionate

zwitterionic compounds

glycine betaine choline-*O*-sulphate

quaternary ammonium compounds

mannitol pinitol

sugar alcohols

Figure 8.2: ***Compatible solutes. The chemical structures of a number of compatible solutes are shown, grouped together as zwitterionic or sugar-based compounds.***

whilst keeping the cytoplasmic concentration of Na^+ and Cl^- ions low (<59 mmol l^{-1}) and storing high concentration of Na^+ (200 mmol l^{-1}) and Cl^- (150 mmol l^{-1}) ions in the vacuole. It is particularly significant that certain crop plants (e.g. rice and tobacco) lack significant amounts of any of these major classes of osmoprotectants. Basis strategies for engineering resistance to water-deficit stress have therefore focused on the production of osmoprotectants as a mechanism for overcoming the osmotic stress generated by water deficit. These strategies have generally required the determination of the biosynthetic pathways for various osmoprotectants, isolation of the relevant genes and appropriate engineering of constructs to target gene expression and protein destination.

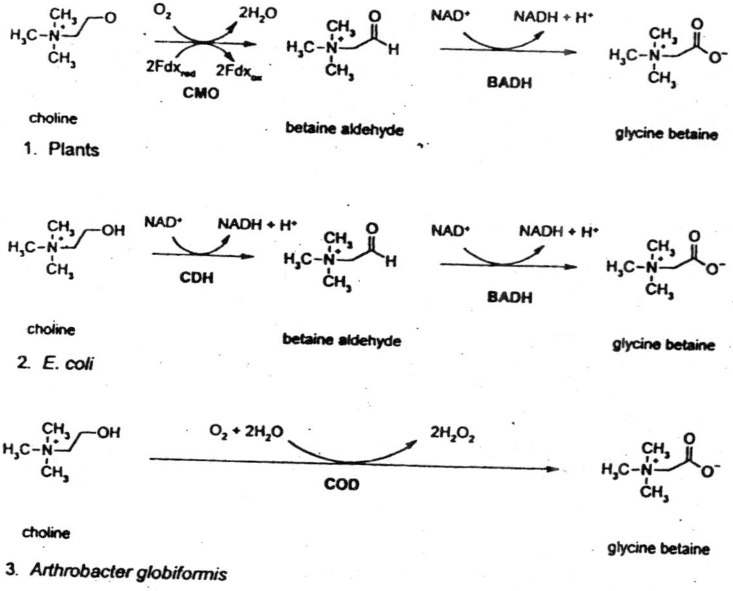

Figure 8.3: ***Pathways to glycine betaine. The different pathways from choline to glycine betaine in plants,*** **E. coli** ***and*** **Arthrobacter globiformis** ***are compared.***

In the same way that production of glycine betaine increases tolerance to a range of stresses that cause water deficit, other osmoprotectants have been produced in transgenic plants to test the feasibility of this approach. Table 8.3 shows the range of genes and products that have been used to test this strategy in a range of crop plants.

What is clear from this table is the existence of a range of effective transgenes for the porduction of osmoprotectants in plants, and that these generally have a measurable effect on the tolerance of the plant to various stresses that causes water deficit. In some cases, tolerance to more than one stress has been demonstrated for the same transgenic plant, confirming the view that genetic strategies to combat water-deficit stress could provide tolerance against a range of different stresses. However, several aspects of

TABLE 8.2 : GLYCINE BETAINE PRODUCTION IN TRANSGENIC PLANTS

Transgene tested	Host plant	Accumulation of glycine betaine	Stress tolerance
Barley *badh* (betaine aldehyde dehydrogenase)	Tobacco peroxisomm	Not tested	Not tested
Spinach *badh*	Tobacco chloroplast	20 μmol g^{-1} FW (in 5 mmol l^{-1} betaine aldehyde)	Not tested
Spinach *cmo* (choline monooxygenase)	Tobacco chloroplast	<0.05 μmol g^{-1} FW	Not tested
E. *coli bet* B (betaine aldheyde dehydrogenase)	Tobacco chloroplast or cytosol	Not tested	Not tested
E. coli bet A (choline dehydrogenase)	Tobacco cytosol (membranes)	Not detected.	Salt
bet A/bet B	Tobacco	0.035 μmol g^{-1} FW	Chilling Salt
bet A	Rice	5.0 μmol g^{-} FW	Drought Salt
A. *glabiformis cod A* (choline oxidase)	*Arabidopsis* chloroplast	1.2 μmol g^{-} FW	Salt Chilling Freezing Heat Strong light
cod A	Rice	5.3 μmol g^{-1} FW	Salt Chilling
A. *pascens cox* (choline oxidase)	*Arabidopsis* Salt	19 μmol g^{-1} DW	Freezing
cox	*Brassica napus*	13 μmol g^{-1} DW	Drought Salt
cox	Tobacco	13 μmol g^{-1} DW	Salt

FW, fresh weight DW, dry weight

osmolyte accumulation still require further investigation and explanation. One point to note from is that the level of accumulation of most of the compatible solutes, apart from the proline and D-ononitol, is considerably lower than could account for their protective effects by osmotic adjustment. The reasons for stress tolerance may be different in each case, with some compounds like glycine betaine acting primarily to stabilise macromolecular structures, whilst other compatible solutes such as mannitol, sorbitol and proline are effective scavengers of reactive oxygen species . Another point to make is that the test of stress tolerance of these transgenic plants are often performed under highly controlled laboratory conditions that present an acute osmotic shock rather than the chronic water-deficit stresses more generally experienced in the field. It should be noted that few of these transgenic plants have been tested for stress tolerance in the field. Another consideration for developing these strategies further is that the accumulation of trehalose and sorbitol was found to inhibit plant growth.

A final note of caution relates to the assumption that increasing osmolyte accumulation in transgenic plants to levels that can maintain cell and tissue turgor pressure under water-deficit conditions will improve crop yield under drought conditions. Under conditions of reduced soil water, the maintenance of leaf turgor and hence transpiration in order to sustain growth will reduce soil water further and hasten the dehydration of the leaves. What needs to be understood is that under conditions where the water deficit is severe enough to threaten plant survival, crop yields are so low that even large increases in fractional yield are of little practical benefit to the grower.

MANIPULATION OF TOLERANCE TO SPECIFIC WATER-DEFIC IT STRESSES

The production of osmoprotectants is not the only possible strategy to engineer resistance to the water-deficit stresses shown in . Individual stresses may be susceptible to specific mechanisms to improve tolerance. Some of these will be described in the following section.

Alternative Approaches to Salt Stress

Salt tolerance is increasingly becoming a major target of crop improvement as substantial areas of irrigated land are damaged by the accumulation of salt. Furthermore, the pressure for land has made it necessary to consider the possibility of growing crops in more saline conditions, with poorer quality water. As described above, saline conditions lead to osmotic stress by preventing water uptake by the roots and water efflux from the cells. However, the accumulation of $Na^+$and Cl^- ions in the cytoplasm may also have direct toxic effects by inhibiting protein synthesis, photosynthesis and susceptible enzymes. Thus, strategies for engineering water stress tolerance via the production of compatible solutes may provide protection against the osmotic effects of saline conditions, but not against ion toxicity. Additional approaches to minimize the toxic effects of specific ions may be required.

Alternative Approaches to Cold Stress

Different plants vary enormously in their ability to withstand cold and freezing temperatures. Most tropical plants have virtually no capacity to survive freezing conditions. On the other hand, many temperate plants can survive a range of freezing temperatures from –5 to –30°C depending upon the species. Plants from colder regions routinely withstand temperatures even lower than this. It is known that plants are better able to withstand cold or freezing stress if they first undergo a period of cold acclimation, at a low but non-freezing temperature. For example, wheat plants grown at normal warm temperature are killed by freezing at –5°C, but after a period of cold acclimation when the plant grows at temperature below 10°C, they can survive freezing temperatures down to –20°C.

Plants differ in their ability to withstand cold or freezing conditions and cold tolerance is one of the traits that plant breeders have selected for over many centuries. However, there has been little improvements in the cold tolerance of major crop species over the past two decades by conventional breeding, prompting the search for molecular solutions to this problem.

Tolerance to Heat Stress

For many years it has been known that heat stress applied to a

TABLE 9.3 : TRANSGENIC PLANTS ENGINEERED TO SYNTHESISE OSMOPROTECTANTS OTHER THAN GLYCINE BETAINE

Osmoprotectant	Transgenes	Crop plants	Accumulation	Stress tolerance
Proline	Mothbean *P5CS*	Tobacco		Salt
	(Pyrroline carboxylate	Rice		Drought, salt, oxidative
	synthetase)	Soybean		Osmotic, heat
	P5CS^{F127A}	Tobacco	4 mg g^{-1} FW	Salt
	(feedback inhibition			
	insensitive)			
	Anti-ProDH	*Arabidopsis*	0.6 mg g^{-1} FW	Salt, freezing
	(proline dehydrogenase)			
Mannitol	*E.coli mt 1 D*	*Arabidopsis*	10 μg g^{-1} FW	Salt
	(mannitol-1-phosphate	Tobacco	6 μmol g^{-1} FW	Salt
	dehydrogenase)			
Sorbitol	Apple *s6pdh*	Tobacco		oxidative stress
	(sorbitol-6-phosphate	Persimmon	61.5 μmol g^{-1} FW	Salt
	dehydrogenase)			

Trehalose	Yeast *tps* 1 (trahalose-6-phosphate synthase, T-6-PS)	Tobacco	3.2 μg g^{-1} DW	Drought
		Potato		Drought
	E. coli ots A + *ots B* (T-6-PS and T-6-P phosphate)	Tobacco	90 μg g^{-1} FW	Drought
D-Ononitol	Ice plant *imtl* (Myo-inositol o-methyltransferase)	Tobacco	35 μmol g^{-1} FW	Drought, salt
Fructans	*B. subtilis sacB*	Tabacco	0.35 mg g^{-1} FW	Drought
		Sugar beet	5 mg g^{-1} DW	Drought
Glutamine	GS2 (chloroplastic glutamine synthetase)	Rice	Salt, chilling	
Osmotin	*Osm1-Osm4* (protein accumulation)	Tobacco	Drought, salt	

FW, fresh weight; DW, dry weight.

wide range of organisms induces a specific set of 'heat stock' proteins (HSPs); they fall into five classes, four of which are highly conserved in prokaryotes and eukaryotes. These four are categorized according to size as the HSP100, HSP90, HSP70 and HSP60 classes whose members appear to function as molecular chaperones. Some of them are expressed constitutively and are involved in normal protein synthesis and folding. Those induced by heat appear to be involved in countering the effects of heat stress by protecting of refolding denatured proteins. Their expressions is induced by heat treatment and, in some cases, can be correlated with the acquisition of thermotolerance. The fifth group of several classes of small heat-shock proteins are particularly abundant in plants, but their function is not yet clear.

In a way analogous to the cold/freezing tolerance strategies, individual heat-shock proteins have been transformed into plants in order to enhance heat tolerance (Table 8.4). However, it is also known that the rapid heat-shock response is co-ordinated by a heat-shock transcription factor (HSF). This protein is expressed constitutively, but in normal conditions exists as a monomer bound to one of the HSP70 heat-shock proteins. Upon heat stress, the HSP70 dissociated and the HSF assembles into a trimer which binds to a heat-shock element (HSE) common to the promoters of heat-shock protein genes. The HSE is made up of 5-bp repeats in alternating orientations with the consensus *NGAAN*; five to seven of these repeats occur in the promoter close to the TATA box. When the AtHSF1 gent was overexpressed in *Arabidopsis*, the transcription factor was not active, and there was no effect on thermotolerance. However, fusion of *AtHSF1* to the N- or C-terminus of the *gusA* reporter gene produced a fusion protein that was able to trimerise in the absence of heat. Transformation of this fusion protein into *Arabidopsis* produced transgenic plants that expressed heat-shock proteins constitutively and demonstrated enhanced thermotolerance without requiring prior heat treatment.

SECONDARY EFFECTS OF ABIOTIC STRESS—

Many, if not all, forms of environmental stress (abiotic and biotic) result in oxidative stress. This occurs most directly as a result of ozone pollution or ionizing radiation. However, oxidative stress is a secondary effect of many type of stress, from pathogen attack to water-deficit stress. The oxidative stress arising from the production of free radicals and the subsequent ensuring cascade of reactions. Much of the oxidative stress-induced damage to cells occurs as a result of the formation of reactive oxygen species (ROS). The cascade of successive reactions that form superoxide, hydrogen peroxide and hydroxyl-radical species. The ROS cause damage as a result of their reactions with cellular macromolecules. Damage to cellular membranes may result from a low concentration of hydroxyl radicals triggering a chain reaction of lipid peroxidation. Oxidative damage to proteins may involves specific amino acid modifications, fragmentation of the peptide chain, aggregation of cross-linked reaction products, altered electrical charge and increased susceptibility to proteolysis. In DNA, both the sugar and bases are susceptible to oxidation by reaction oxygen species, causing base degradation, single-strand breakage, and cross-linking to protein. These lesions in DNA can result in deletions, mutations and other damaging genetic effects.

Plants contain a number of enzymes that catalyse the cascade of reactive oxygen species and convert them to less reactive products. One key point of notice is the key role of certain enzymes in this cascade, notably superoxide dismutase, catalase and peroxidases in neutralizing these reactive species. Also shown in the figure are a number of 'antioxidant' compounds that react with the activated oxygen compounds to produce harmless, regenerable products. These include three different vitamin compounds: β-carotene (provitamin A), ascorbic acid (vitamin C) and α-tocopherol (vitamin E). Other important antioxidants include glutathione and zeaxanthin. Central to this antioxidant system is the ascorbate-glutathione cycle. The coupling of ascorbate and glutathione redox-cycling has been most extensively characterized in the chloroplast. Chloroplasts produce superoxide and hydrogen peroxide in the light, particularly from photosystem I. The superoxide generated is itself converted into hydrogen peroxide by either spontaneous dismutation or by the activity of superoxide dismutase. The hydrogen peroxide produced is scavenged by ascorbate and the enzyme ascorbate peroxidase. The monodehydroascorbate produced from this reaction may be regenerated in two ways. One is by reduction of NAD(P)H catalysed by monodehydroascorbate reductase. The other occurs via the dismutation of two monodehydroascorbate molecules to form ascorbate and dehydroascorbat, followed by the reduction of the dehydroascorbate by glutathione, catalyzed by dehydroascorbate reductase.

From the outline of plant antioxidants and oxygen-scavenging activities given above, two general strategies for engineering tolerance to oxidative stresses are apparent—either to increase the level of enzymes that remove ROS, or to increase the level of antioxidant compounds that react with ROS.

TABLE 8.4 : TRANSGENES USED TO MANIPULATE HEAT TOLERANCE

Gene	Protein	Transgenic plant
AtHSF1	Heat-shock transcription factor HSF1 **GUS fusion	*Aradibopsis*
Hsp101	HSP100 class heat-shock protein	*Arabidopsis*
Hsp70	HSP70 class heat-shock protein	*Arabidopsis*
Hsp 17.7	SmHSP small heat-shock protein family	Carrot
TLHS1	Class 1 smHSP	Tobacco

Strategy 1: Expression of enzymes involved in scavenging ROS

Superoxide dismutases (SODs) catalyse the reaction shown in Figure 9.4, in which the superoxide radical combines with two hydrogen ions to form oxygen and hydrogen peroxide. The activity of this enzyme therefore determines the concentration of the two substrates (superoxide and hydrogen peroxide) that react to form the hydroxyl radical. There is a broad family of SOD enzymes, requiring different metal ions for activity, and predominating in different cellular compartments. Thus, Mn⁻, Fe⁻ and Cu/Zn⁻ forms are found, specific to different organelles.

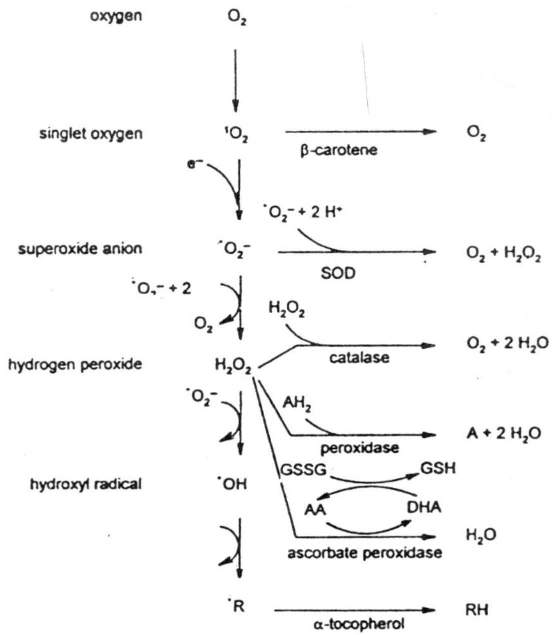

Figure 8.4: ***Formation of ROS and enzymes that remove ROS. The formation of reactive oxygen species and the cascade of subsequent reactions that result are shown. The roles of the antioxidants and antioxidant enzymes are shown. AA, ascorbic acid; DHA, dehydroascorbate; GSH, glutathione; GSSG, oxidised gluthatihone; SOD, superoxide dismutase.***

Transgenic tobacco plants containing each class of enzyme have been produced and tested for their ability to withstand oxidative stress, as shown in. In each case, there has been some measurable increase in the ability to withstand oxidative stresses such as exposure to ozone. The Mn-SOD has also been transformed into alfalfa and targeted to the chloroplast or mitochondria. In this case, the results of a 3-year field trial indicated significant improvements in the yield and survival of the transgenic plants. However, it is not clear whether the protective effect is due to the removal of superoxide by the enzymes, or whether the hydrogen peroxide product enhances stress tolerance by inducing other stress-related genes. More recently, the introduction of Cu/Zn=SOD, Mn-SOD and Fe-SOD genes together into alfalfa has been shown to increase the winter hardiness of the crop.

TABLE 8.5 : TRANSGENES USED TO ENGINEER TO OXIDATIVE STRESS

Gene	Host	Stress tolerance
Mitochondrial Mn-SOD Tobacco	Alfalfa chloroplast	2× increase in SOD Increased field drought tolerance
Mitochondrial Mn-SOD	Tobacco chloroplast	2-4× increases ozone tolerance
Mitochondrial Mn-SOD Tobacco	Tobacco mitochondria	8× increase in SOD No effect on ozone tolerance
Mn SOD	Canola	Increased aluminium tolerance
Chloroplast Cu/Zn-SOD Pea	Tobacco chloroplast	3-15× increases in SOD Increased tolerance to high light and chilling
Cytosolic Cu/Zn-SOD	Tobacco cytosol	1.5-6× increases in SOD Reduced damage from acute ozone exposure
Fe-SOD *Arabidopsis*	Tobacco	Protected plants from ozone damage
Apx 3 (ascorbate peroxidase)	Tobacco	Increased protection against oxidative stress
Apx I (ascorbate peroxidase)	*Arabidopsis*	Heat tolerance
GST/GPX (glutathione S-tranferase with gluthtathione peroxidase)	Tobacco	Increased stress tolerance
Nt107 (glutathione S-transfearse)	Tobacco cold and salinity stress	Sustained growth under
Par B (glutathione S-transfearse)	*Arabidopsis*	Protects against aluminium toxicity and oxidative stress
Nt Pox (glutathione peroxidase)	*Arabidopsis*	Protects against alu-

		minium toxicity and oxidative stress
Glutathione reductase *E. coli*	Tobacco chloroplast	3-6× increase in foliar Gr Increased tolerance to SO_2 and paraquat
Glutathione reductase *E.coli*	Tobacco cytosol	1-35× increase in GR Increased tolerance to paraquat
Glutathione reductase *E.coli*	Poplar cytosol	100× increases in GS GSH not increased No effect on paraquat tolerance
MsFer	Tobacco	Increased tolerance to Alfalfa feritin oxidative damage caused by excess iron

SOD, superoxide dismutase; GST, glutathione S-transferase; GPX, glutathione peroxidase; GR, glutathione reductase; GS, glutathione synthetase; GSH, glutathione.

Strategy 2: Production of antioxidants

Also shows the alternative strategy for oxidatives stress tolerance in action. Thus, three of the enzymes—ascorbate peroxidase, glutathione peroxidase and glutathione reductase—have been transformed into *Arabidopsis* and tobacco plants and shown to have some effect on tolerance to various abiotic stresses such as heat, cold and salinity. Glutathione reductase also provided resistance to the oxidative stress resulting from paraquat treatment. On the other hand, simply increasing the cellular concentration of glutathione by expressing glutathione synthase had no effect on stress tolerance.

9

Molecular Farming

Molecular farming' is a term coined to describe the application of molecular biological techniques to the synthesis of commercial products in plants. The term can be applied to a broad spectrum of activities, from the enhanced production of products that are already extracted from plants through to the manufacture of compounds that are completely novel to plants. A wide range of products have already been identified as likely targets for molecular farming, these include a variety of carbodydrates, fats and proteins as well as secondary products. This chapter will consider examples of some of the major classes of molecules produced in transgenic plants. The first part of this chapter will look at some examples of molecular farming as applied to carbohydrates and oils. In the second part of the chapter, examples of molecular farming in which it is the transgenic protein itself that is the product will be looked at. Many of the proteins being 'farmed' in plants are antibodies, vaccines or biopharmaceuticals aimed at improving human and animal health, and it is in the area of molecular farming that many of the most exciting and potentially beneficial developments in plant biotechnology are taking place. At the end of the chapter, we will also review some of the economic reasons why molecular farming might, in some cases, be an attractive alternatives to current forms of manufacture of these compounds.

TABLE 9.1 : MAJOR TARGETS FOR CARBOHYDRATE AND LIPID MOLECULAR FARMING

Compound	Origin of genes	Application	Transgenic plant
Carbohydrates			
Amylose-free starch	Potato	Food, industrial	Potato
High amylose starch	Potato	Food, industrial	Potato
Cyclodextrins	*Klebsiella pneumonia*	Food, pharmaceutical	Potato
Fructans	*Bacillus subtilis*	Industrial	Tobacco, potato, maize
jerusalem artichoke	Food	Sugar beet	
Trehalose	*Escherinchia coli*	Food	Tobacco
Lipids			
Medium-chain fatty acids	Calinornia bay tree	Food, detergent, industrial	Oilseed rape
Saturated fatty acids	*Brassica rapa*	Food	Oilseed rape
Mono-unsaturated fatty acids	Rat	Food	Tobacco
Polyhydoxybutyrate	*Alcaligenes eutrophus*	Biodegradeable plastics	*Arabidopis*, soyabean, Oilseed rape, cotton

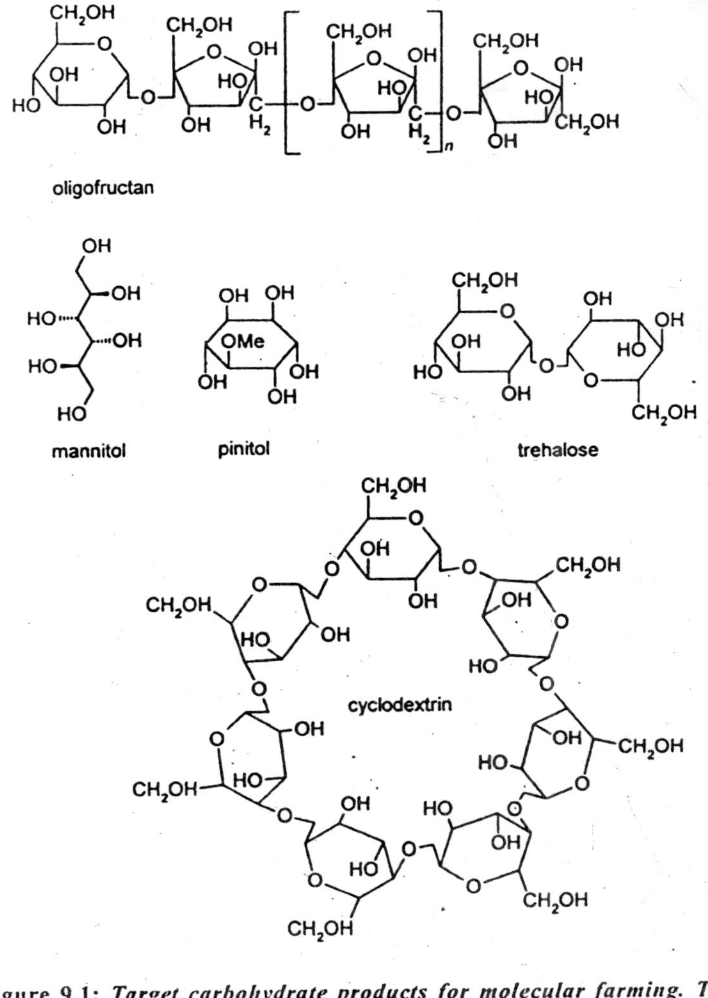

Figure 9.1: ***Target carbohydrate products for molecular farming. The structures of some potential targets for molecular farming are shown. Oligofructans are polymers produced by the sequential addition of fructose on to an initial sucrose molecule. The sugar alcohols mannitoal and pinitol were discussed in Chapter 9 with reference to their roles as oscolytes, and the potential for producing them in crops to enhance stress tolerance. Trehalose is a disaccharide with useful properties as a food preservative. Cyclodextrins are rings or 6, 7 or 8 glucose residues. The 7-sided ring shown here has a number of pharmaceutical application for the solubilasation of small hydrophobic molecules.***

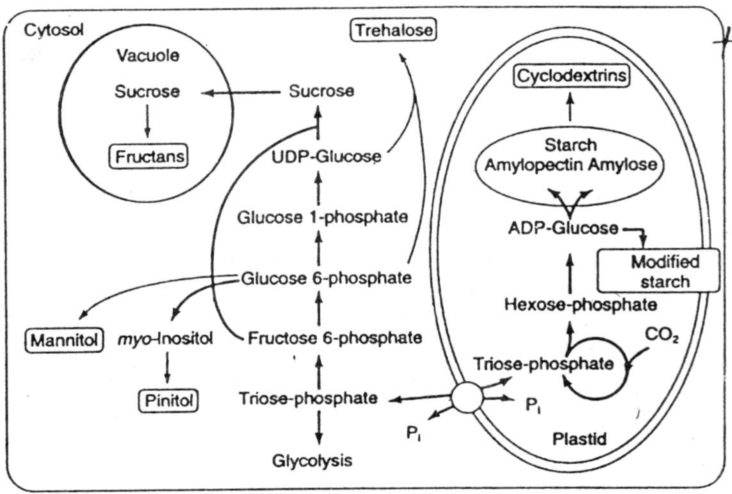

Figure 9.2: ***Metabolic pathways for the biosynthesis of carbohydrate products for molecular farming. The biosynthetic pathways for the production of the compounds shown in are shown schematically in this stylised cell. Starting from the pool of hexose-phosphate sugars in the cytoplasm, sucrose is synthesised via UDP-glucose and transported into the vacuole, where it may be stored as fructans. UDP-glucose and glucose 6-phosphate are used to form trehalose in the cytoplasm, and glucose 6-phosphate is also the precursor of the sugar alcohols. In leaf cells, triose-phosphates are transported in and out of the chloroplast, where they can be converted to hexose-phosphate and then ADP-glucose for starch synthesis. Cyclodextrins are not found in plants, bu. starch is a substrate for some bacterial biosynthetic enzymes.***

CARBOHYDRATES AND LIPIDS

In this first section we will look at some examples (see Table 9.1 for a summary) of the production of novel or modified carbohydrates and oils (and their derivatives) in plants. Many of these examples have multiplied applications (both as modified food- or feedstuffs and industrially) that will be highlighted here.

Carbohydrate Production

Plants produce a range of commercially valuable carbohydrates. The two most abundant carbohydrates are cellulose (used for fibres from cotton and flax, for paper-making from trees and for a range of industrial products such as paints and polymers) and starch (used for food, feed and industrial purposes). Some biotechnological effect is directed towards improving the yield and quality of these bulk carbohydrates. However, there are a number of other carbohydrates that it could be attractive to produce in transgenic plants, including oligofructan, cyclodextrins and trehalose.

Indicates the biosynthetic pathways and cellular locations of these compounds in plants. One obvious point is that carbohydrates are synthesized and stored in a number of different cellular compartments. Thus, starch and its derivatives are synthesized in plastids, whilst the fructans, another important class of storage carbohydrates in plants, are synthesized and stored in the vacuole. Sugars and sugar alcohols are synthesized in the cytosol and accumulate throughout the cell, often in response to abiotic stress. Note also that a stylised cell, and that in practice different carbohydrates tend to accumulate in different tissues and organs. For example, starch is stored transiently in the leaves of cercals, and starch reserves accumulate in the amyloplasts of the grain endosperm. However, the fructans are also a significant storage carbohydrate in some cereals, and accumulate in the leaves and stems.

Cyclodextrins from starch

Whilst starch itself is a bulk feedstock for industry, the potential exists for carrying out further biotransformations on the starch in the plant, rather than by chemical or fermentation processes after extraction. One type of high-value product that could be made from starch is the cyclodextrins. These compounds are typically 6-, 7- or 8-membered rings comprising glucopyranose subunits attached in $\alpha(1 \rightarrow 4)$ linkages. These compounds are normally produced by bacterial fermentation of maize starch, particularly for pharmaceutical applications. The structure of the cyclodextrins is such that they form a cone-shaped ring, with the hydrophilic residues on the exterior and a hydrophobic pocket in the centre of the ring. The 7-membered ring has the ideal dimensions to form a pocket for small hydrophobic compounds. Thus, in a concentrated suspension, the cyclodextrins will effectively solubilize hydrophobic pharmaceuticals such as steroids. At lower concentrations, for example after injection into the bloodstream, the therapeutic agent is released.

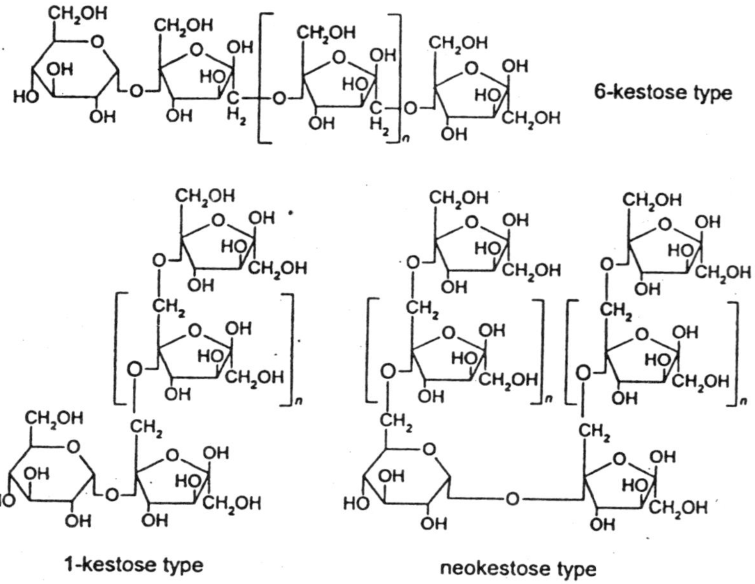

Figure 9.3: ***Structure and synthesis of polyfructans. The structures of three types of polyfructan are shown. The levan-type fructans of the 6-kestone class are formed by fructose (6 → 2β)-fructose linkages. These fructans are typically found in grasses, with n = ~200. (In bacteria, levans are much are larger polymers than this). The 1-kestose type of fructans are also called inulins, and are formed by fructose-(1 → 2β)-fructose linkages. These are typically found in bulbs and the storage roots of plants such as jerusalem artichoke and chicory, with n < 50. Fructans of the neokestose type, with 1-kestone chains attached to both the glucose and fructose residues of the initial sucrose moiety, are the smallest fructans, with only 5-10 fructose residues.***

Having been identified as a suitable target for molecular farming, it is striking that only one report, in 1991, of an attempt to produce cyclodextrins has been published. A bacterial cyclodextrin glycosyltransferase gene from *Klebsiella pneumoniae* was fused to a plastid-targeting sequence and placed under the control of the promoter from the patatin gene. Patatin is a protein that accumulates in potato tubers, and the promoter directs high levels of expression in the tubers of transgenic potatoes. However, transformation of potatoes with this construct resulted in very little conversion (0.001-0.01%) of starch to cyclodextrins. It was concluded that the insoluble starch granules may have been inaccessible to the bacterial enzyme, or that the enzyme granules may have been inaccessible to the bacterial enzyme, or that the enzyme became trapped in the growing granule. Whatever the reason, no subsequent attempts to produce this commodity have appeared in the literature.

Figure 9.4: ***Consturct for the synthesis of fructans in the vacuole of transgenic plants. A construct of the type used transform tobacco in the first experiments to engineer fructan synthesis is shown. The bacterial* sacB *gene is flanked by an N-terminal vacuolar targeting sequence from yeast carboxypeptidase. Note the tandem-repeated CaMV 35S promoter and AIMV translational enhancer to obtain high levels of expression.***

Trehalose

Trehalose is produced in some plants and microorganisms (e.g. yeasts), often in response to osmotic stress. It is therefore another potential target for the genetic manipulation of tolerance to abiotic stresses that create water deficit. However, it is also a valuable commodity for food processing, dehydration and flavour retention. Genes for trehalose synthesis from yeast and *E. coli* have already been introduced into transgenic tobacco, but the purpose of these experiments was to manipulate drought tolerance rather than to manufacture bulk quantities of the sugar.

Metabolic Engineering of Lipids

The next class of compounds that we will explore as targets for molecular farming is the lipids and their derivatives. As with the carbohydrats, lipids are already produced in large quantities from major crops such as Oilseed rape (canola), soybean and maize, for industrial as well as food purposes. Molecular farming will therefore have a role in the same broad areas as shown for carbohydrates, that is:

- improvement of existing lipid products;
- engineering of novel lipid products.

Improvement of plant oils

In daily life, our closest association with plant oils comes by way of their rapid displacement of animals fats as cooking oils. This association alone should indicated that oils are already extracted from a number of different types of crop. Oilseed rape or canola, and soybean are major sources of oil that are often unspecified, but sunflower oil, corn oil and olive oil are often identified because of the improved characteristics of these oils for particular purposes. Plant oils are also used in a wide range of processed foods, and not for animal feeds. They also have an increasing number of applications in industry for the manufacture of soaps and detergents, lubricants and biofuel. However, the non-food applications currently comprise only about 10% of the total vegetable oils produced. One of the reasons for this relatively low level of industrial use is the complex and variable nature of plant oils, which generally makes them less suitable for oleochemical application then cheaper alternatives from mineral oils. A major long-term goal of this sector is therefore the improvement of plant oils for industrial applications, since they must eventually replace non-renewable, petroleum-based products.

Storage oils that are used as carbon/energy reserves in plant seeds and some fruits (such as olives and avocados) are generally glycerol esters of fatty acids, called triacylglycerols (TAGs) . The differences between oils from different sources are largely due to the particular proportions of different fatty acids comprising TAG. A few of the common fatty acids are also shown in. These differ in their length, and the number and positions of double bonds. Saturated fats contain no double bonds—the more double bonds there are, the greater the degree of unsaturation.

The site of *de novo* fatty acid biosynthesis is exclusively in the stroma of the plastid, whereas most modification of the fatty acids occurs in the cytoplasm and on the endoplasmic reticulum (ER). Assembly of TAGs takes place in the membrane of the ER. The TAGs are stored in oil bodies, which are essentially oil droplets surrounded by a lipid monolayer formed from one half of the ER membrane lipid bilayer. The oil bodies of seeds (and other tissues that undergo extreme desiccation) also contain proteins, called 'oleosins', in the lipid monolayer.

The first committed step in fatty acid biosynthesis in the plastid is the formation of malonyl-coenzyme A (CoA) from acetyl-CoA, in an ATP-dependent reaction catalysed by acetyl-CoA carboxylase (ACCase). The CoA residue is then exchange for the acyl carrier protein (ACP). Acetyl-CoA is then condensed to the malonly-ACP, forming acetoacetate-ACP, and liberating CO_2. Three subsequent steps (reduction of the carbonyl group, removal of water to form a double bond and reduction of the double bond) produce and acyl-ACP, which is two carbons longer than the original. The entire sequence of elongation reactions from the initial binding to ACP is catalysed by a fatty acid synthase multienzyme complex. The elongated fatty acid chain is then transferred to another ACP protein and this is then condensed to a new malonyl-ACP. Thus, the formation of fatty acids occurs by the stepwise addition of 2-carbon units at the carboxyl end, hence the even numbers of carbon atoms in the fatty acids shown in. In many of the oil producing crops, this process stops at the 16-carbon stage,and the palmitoyl-ACP (16:0) is elongated to stearoyl-ACP (18:0) by a specific synthase. (The standard nomenclature of fatty acids indicates the number of carbon atoms and the number of double bonds). At this stage, desaturation by a soluble Δ^9C-stearoyl desaturase in the plastid stroma converts most of the stearoyl-ACP to oleoyl-ACP (18:1Δ^9). (The Δ^9 indicates that the double bond starts at the 9th carbon counting from the carboxylic acid 'front end' of the molecules.)

After termination of synthesis in the plastid, the fatty acids (mainly palmitic, strearic and oleic acids) are released from the ACP and exported to the cytoplasm, where they are converted to acyl-CoA esters. TAGs are formed by stepwise acylation of glycerol-3-phosphate in the ER membrane. Further modifications to the fatty acids normally occur after they have been attached to various glycerophosphatides. Additional double bonds may be added (typically to Δ^{12}, Δ^{15} and Δ^6 as well as Δ^9). Other modifications may include hydroxylation by the addition of water across a double bond.

Fatty acids

Palmitic acid (16:0)

Slearic acid (18:0)

Oleic acid (18:1, Δ^{9c})

Linoleic acid (18:2, $\Delta^{9,12c}$)

α-Linoleic acid (18:3, $\Delta^{9,12,15c}$)

Erucic acid (22:1, Δ^{13}-*cis*)

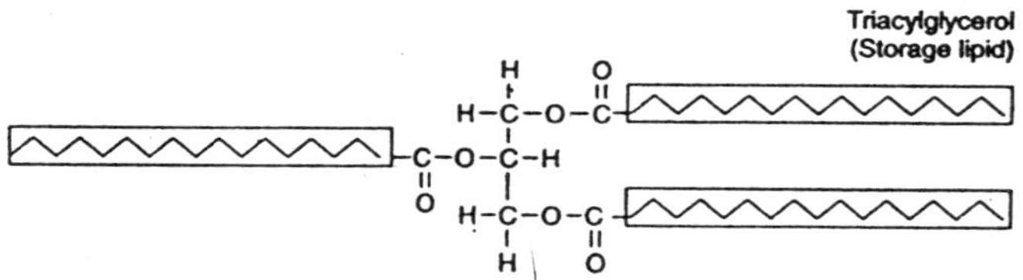

Figure 9.5: *Fatty acid and triacylglycerol structure. The structures of some of the common fatty acids found in plants oils are shown. The standard nomenclature in brackets denotes the number of carbon atoms, the number of double bonds and the position of the double bonds, counting from the carboxyl-, or Δ-end. Thus, linoleic acid (18:2$\Delta^{9,\ 12}$) has 18 carbon and 2 double bonds, one form C9-C10, and the other form C12-C13. The structure of a triacylglycerol storage lipid is also shown, indicating the attachment of three fatty acid chains to a glycerol molecule. Note that apart from the glycerol residue, these molecules are completely hydrophobic—compare this structure with the phospholipids of biological membranes, where one of the fatty acids is replaced by a polar 'head' group.*

TABLE 9.2 : THE FATTLY ACID COMPOSITION OF MAJOR OIL CROPS

Fatty acid	Soya oil	Palm oil	Rape-seed oil (LEAR)	Sunflower oil	Peanut oil
16:0	11	42	4	5	10
18:0	3	5	1	1	3
18:1	22	41	60	15	50
18:2	55	10	20	79	30
18:3	8	0	9	0	0
20:1	0	0	2	0	3
Others	1	0	2	0	0

LEAR, low erucic acid rape.

Production of shorter chain fatty acids

The oils produced by the major oil crops of the world consist mainly of palmitic, stearic, oleic, linoleic and linolenic acids, which are all C16 or C18 in length. Coconut and palm kernel oils are largely C8-C14, and lauric acid (12:0), in particular, is an important raw material for the production of soaps, cosmetics and detergents (think of the widespread use of sodium lauryl sulphate, or SDS, in these products). The synthesis of fatty acids at a particular length is terminated by the hydrolysis of the acyl-ACP by a thioesterase. The California bay tree contains a very high proportion of lauric acid in its seeds, and an acyl-ACP thioesterase that specifically hydrolyses lauroyl-ACP has been cloned from this source. The introduction of this gene into oilseed rape causes fatty acid synthesis to terminate at the 12:0 stage, and a high proportion of lauric acid to accumulate in the seed oil. Most importantly, field tests show that these plants grow normally and produce normal yields

Production of longer chain fatty acids

One of the important targets is to produce fatty acids longer than C18 for use as industrial oils. In oilseed rape and other *Brassica* species, there is a two-step elongation pathway from oleoyl-CoA ($18 : 1\Delta^{9}$) to erucoyl-CoA ($22 : 1\Delta^{13}$), such that erucic acid is one of the constituents of brassica oils. However, whilst erucic acid is valuable as an industrial oleochemical, it is nutritionally unsuitable for human consumption. Conventional breeding has led to the development of two distinct Oilseed rape crops—high-erucic acid rape (HEAR) for industrial purposes, and low-erucic acid rape (LEAR) with virtually no erucic acid, for food products (see . (NB: the existence of two distinct crops that need to be kept apart means that systems are in place in the seed production, agriculture and processing industries to ensure minimal cross-contamination, a process called 'identity preservation'. Thus, requirements to keep GM and non-GM seeds and products separate do not create entirely new demands on this sector). However, the highest erucic acid content of HEAR is about 50% of the total fatty acids, which makes the cost of separating out and disposing of the other acids uncompetitive with mineral oil sources. Attempts are being made to overexpress the genes that encode the elongases, and also to transfer enzyme activities that preferentially incorporate erucic acid into TAGs.

Modification of the degree of saturation

As the erucic acid example shows, the ideal products for industrial purposes are as uniform as possible. This requires uniformity of desaturation as well as length. (The same does not necessarily apply to food constituents, where complex mixtures may produce the ideal product. In the case it is consistency, rather than uniformity, that is required). One of the first attempts in this area was to increase the level of stearic acid (18:0) at the expense of oleic acid (18:1Δ^9) by inserting an antisense construct of the Δ^9 desaturase gene into oilseed rape. The antisense gene was under the control of the napin (a seed-specific protein of brassica) gene promoter. There was a marked decrease in the amount of desaturase enzyme, resulting in a decreased formation of oleic acid, and a rise in the proportion of stearic acid from 1-2% up to 40% of total fatty acids. This high stearic acid oil has potential as a cocoa butter substitute.

On the other hand, there is considerable potential in the production of a very high (>90%) oleic acid oil for food purposes and as a uniform oleochemical feed-stock. Conventional breeding has produced mutal oilseed rape lines with an oleic acid content close to 80%, but attempts to increase this level have produced plants with poor cold tolerance, presumably as a result of the lack of unsaturated fatty acids in the cellular membranes. Antisense repression and co-suppression of the Δ^{12} desaturase in the seeds of oilseed rape have, however, facilitated the raising of oleic acid levels to 87-88%, without affecting cold tolerance. Similar experiments in soybean increased the oleic acid level from 22% to 79%.

Production of rare fatty acids

There are 210 known types of fatty acids produced in plants, but most of these are not found in the major crops plants and would be difficult to produce commercially in their host plants. However, it is possible to transfer genes from these plants to manipulate the profile of fatty acids produced in the major oil crop plants. One such fatty acid is petroselenic acid ($18{:}1\Delta^6$), which is found in certain Umbelliferae, such as coriander, and has potential as a raw material for industry. Oxidation of petroselenic acid by ozone produces lauric acid (12:0) (for soaps and detergents) and adipic acid (6:0), which can be used for nylon production. Transformation of tobacco with a coriander acyl-ACP desaturase cDNA led to the production of petroselenic acid in calli to a level of 5% of total fatty acids.

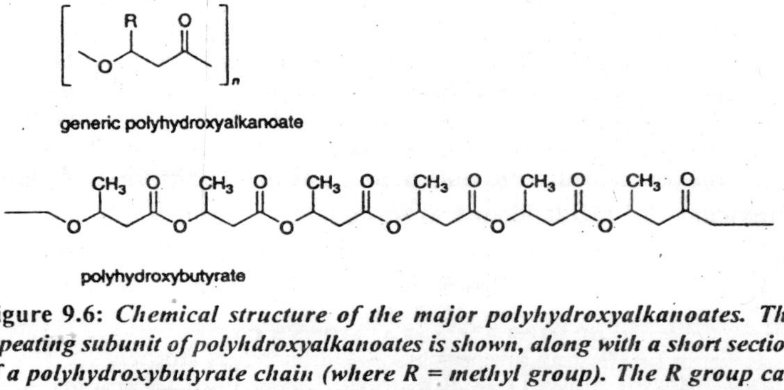

Figure 9.6: ***Chemical structure of the major polyhydroxyalkanoates. The repeating subunit of polyhdroxyalkanoates is shown, along with a short section of a polyhydroxybutyrate chain (where R = methyl group). The R group can be anything from 0 carbons (i.e. H) in hydroxypropionate, to > 10. The length of the R-group side chain affects the physical properties of the plastic formed by these polymers.***

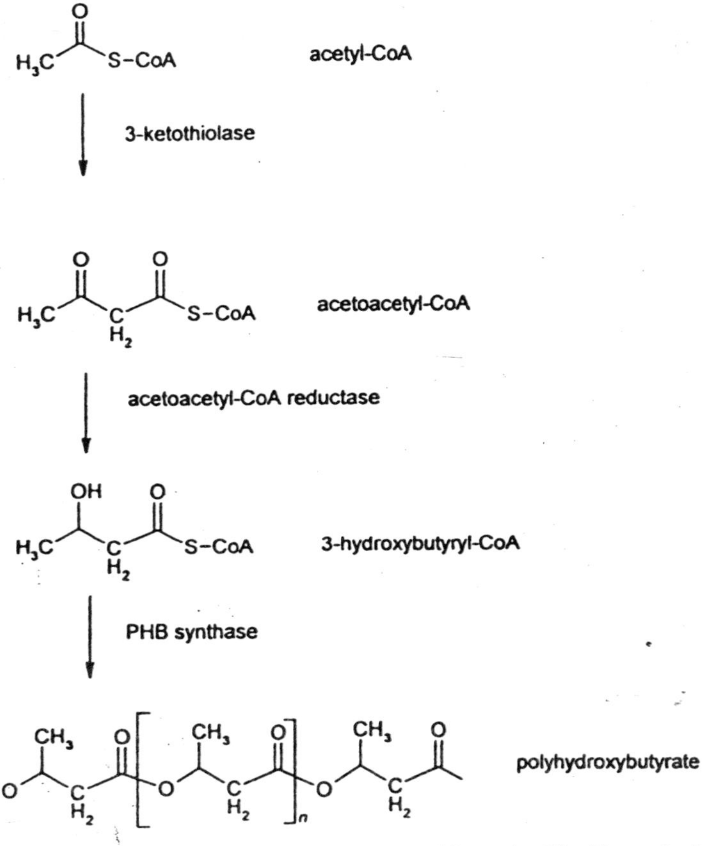

acetyl-CoA

3-ketothiolase

acetoacetyl-CoA

acetoacetyl-CoA reductase

3-hydroxybutyryl-CoA

PHB synthase

polyhydroxybutyrate

Figure 9.7: ***The PHB biosynthetic pathway of bacteria. The biosynthesis of polyhydroxyburyrate requires only three steps from acetyl-CoA. In the first step, two acetyl-CoA residues are condensed by 3-ketothiolase to form acetoacetyl-CoA. This is reduced by acetoacetyl-CoA reductase to give 3-hydroxybutyryl-CoA. In the final stage, this compound is polymerised by PHB synthase***

Certain polyunsaturated fatty acids have pharmaceutical or nutraceutical value. These include γ-linolenic acid ($18{:}3\Delta^{6,9,12}$) and arachidonic acid ($20{:}4\Delta^{5,8,11,14}$), which are essential fatty acids for humans and precursors of eicosanoids (including prostaglandins, leukotrienes and thromboxanes). Recently, it has become clear that the synthesis of these compounds involves a specific subclass of microsomal desaturases called 'front-end'

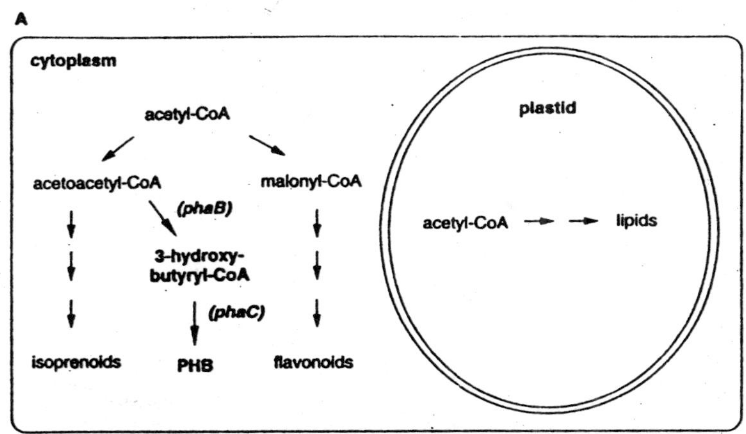

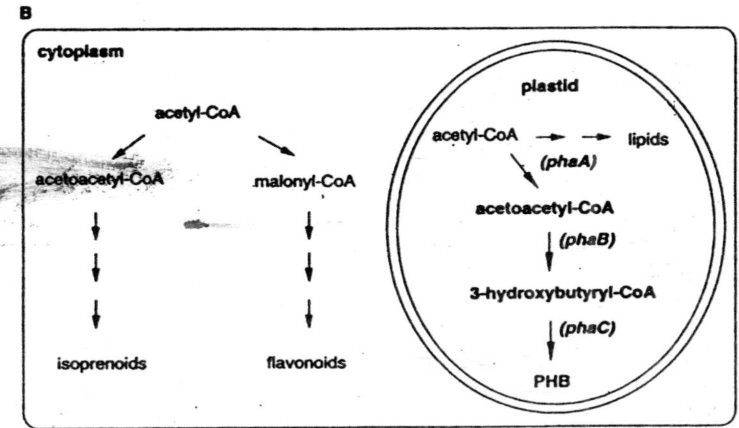

Figure 9.8: ***Expression of the PHB biosynthesis pathway in transgenic*** **Arabidopsis.** ***In scheme A, the presence of 3-ketothiolase activity in the plant cytoplasm removes the necessity to transfer three genes to the plant. The*** **phaB** ***and*** **phaC** ***genes from*** **Alcaligenes eutophus** ***were transformed into*** **Arabidoposis** ***without protein sequences. The PHB granules were produced in the cytoplasm, but they also accumulated in the nucleus and vacuole. In scheme B, all three genes of the pathway are targeted to the chloroplast using Rubisco small-subunit transit peptide sequences. In this case, PHB accumulates in the plastids.***

desaturases. These enzymes insert additional double bonds between existing bonds and the carboxyl 'front-end' of the fatty acid, whereas most desaturases in plants add sequentially toward the methyl end. The transformation of a front-end desaturase from borage (which is one of a few plant species to produce γ-linolenic acid) into tobacco resulted in the accumulation of high levels of Δ^6 unsaturated fatty acids. A Δ^5 desaturase has been cloned from the filamentous fungus *Mortierella alpina* (which accumulates arachidonic acid) and expression in oilseed rape produced significant levels of polyunsaturated fatty acids.

Ricolenic acid (Δ^{12}-hydroxyoleic acid) is produced in castor beans to a level of 90% of the total fatty acids. However, the castor oil crop has a number of problems, including the presence of toxic compounds such as ricin in the residual meal. The synthesis of ricolenic acid involves the direct hydroxylation of oleic acid bound to phosphatidylcholine on the ER membrane. The cDNA for the 12-hydrocylase has been cloned from castor bean and could be used to produce 'castor oil' in major oil crops. Another fatty acid modifying enzyme has been cloned from *Creptis acetylenics*. Thise non-haem, di-iron protein catalyses triple bond and epoxy group formation in fatty acids, and would be a valuable way of inserting chemically reactive sites into oils

MOLECULAR FARMING OF PROTEIN

Throughout this book we have presented to you examples where transgene expression is used as a tool to, for example, modify a biosynthetic pathway, manipulated growth and development of alter some other property of the plant as herbicide resistance. However, in some cases it is the transgenic protein itself that is important. We see how attempts to improve the amino acid content of grain are being addressed by expression of proteins with a high content of lysine. In this section of the chapter, we will consider the production of bulk enzymes and potentially high-value proteins such as antibodies and vaccines. The production of functional proteins in plants on an industrial scale does not, as first sight, suggest too many diffculties. However, the production of a complex protein that is folded, processed, easily purified and can be shown to be safe for pharaceutical use, all at low cost, is a real challenge. The economics of protein production in plants is complicated. The actual cost will depend on many factors, amongst them being the cost of growing the plant, transport costs and processing and protein purification costs. The costs of protein produced in plants may undercut the costs of producing proteins by existing methods, but if too much (in economic terms) is produced, then the market price will fall. This has already proved to be the case for some vaccines produced by existing methods.

In this remaining section, the protein products that will be considered fall into two main types. First, enzymes for industrial and agricultural uses will be looked at. Second, we will consider medically related proteins such as:

- antibodies—full immunoglobulins and engineered types such as scFvs (single chain antibodies);
- subunit vaccines—vaccines based upon short peptide sequence that act as antigenes rather than whole organism-based vaccines; and
- protein antibodies.

Production Systems

The interest in producing such proteins in plants comes in part from the problems associated with existing fermentation or bioreactor systems. Mammalian cell systems are expensive and cannot easily be scaled up (input costs can approach $US1 million per kg of crude product). Bacterial systems can be scaled up, but often the recombinant proteins are not properly processed (they are not properly folded and disulphide bridges are not formed), so intracellular precipitation of non-functional proteins can occur. Plant systems can be scaled up so that large amounts of material can be harvested and processed, allowing industrial-scale amounts of protein to be purified. In some cases it may be possible to omit purification, as is the case with vaccines from plant source, and where plant material containing recombinant enzymes can be added directly to animal feed or industrial processes. There are some differenece between plant and animal post-translational modification systems, and the difference in glycosylation pattern between plants and animals has caused some concerns, particularly when expressing antibodies in plants. However, plants are able to fold, cross-link, and post-translationally modify (by, for example, glycosylation) non-plant protein sufficiently well to ensure that, in most cases, functional proteins are obtained. The production of recombinant protein in plants also benefits from the ability to target the proteins to cellular compartments where processing occurs, or where the proteins may be more stable. Another advantage of using plant systems is that the potential for contamination of the protein products with toxins or pathogens of animals and humans is very much reduced.

TABLE 9.3 : POTENTIAL BIOMASS YIELDS FROM VARIOUS PLANTS CONSIDERED AS POTENTIAL VEHICLES FOR PROTEIN PRODUCTION

Crop	Potential annual biomass yield (tonne per hectare)
Tobacco	>1000
Alfalfa	25
Maize	12
Rice	6
Wheat	–3

The plants initially chosen for expression studies related to the ability of scientists to transform the plant material. A lot of work was first done with tobacco, but this was not a suitable plant for feeding to animals. For work in which edible products are being developed, plants such as potatoes, tomato, maize and lettuce have been used. For bulk production, the leaves of plants such as tobacco and alfalfa remain a favourite, because they can be harvested several times a year. However, plants such as rice, wheat, maize and soybean have also been used, although their biomass yields are lower.

Strategies for protein production

Two major strategies have been developed for the production of proteins in plants: the stable integration approach, which has been used for many systems; and the use of plant viruses as transient vectors.

Stable expression

The stable transgene expression approach, in which the transgene is regulated by a strong, constitutive promoter (such as the 35S promoter), is perhaps the most suitable for the bulk production of soluble proteins in leaves, although yields can be low using this approach. A more sophisticated approach has been to

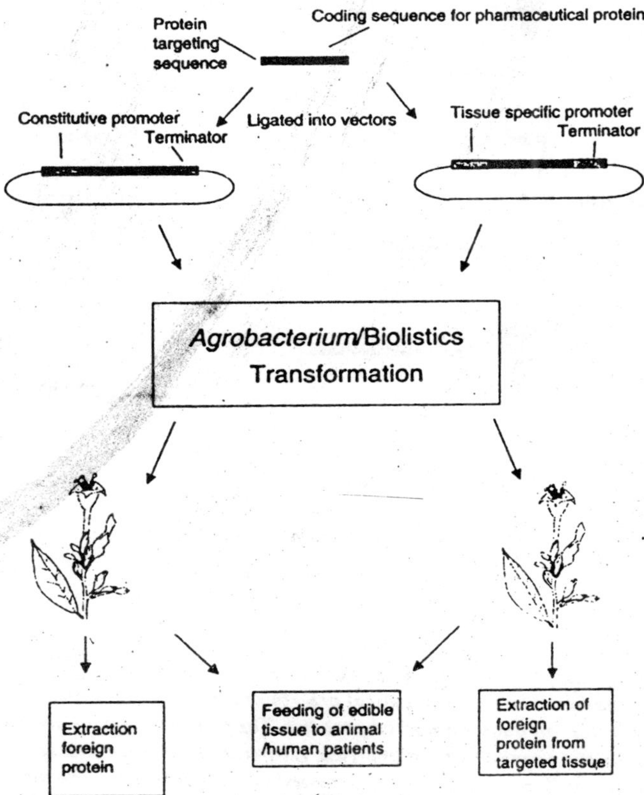

Figure 9.9: ***Different routes of protein manufacture in stable transformed plants. Constitutive (from, for instance, the 35S promoter) or targeted (to particular organs or tissues) expression and accumulation results in recombinant proteins that can either be used after purification or directly as plant products.***

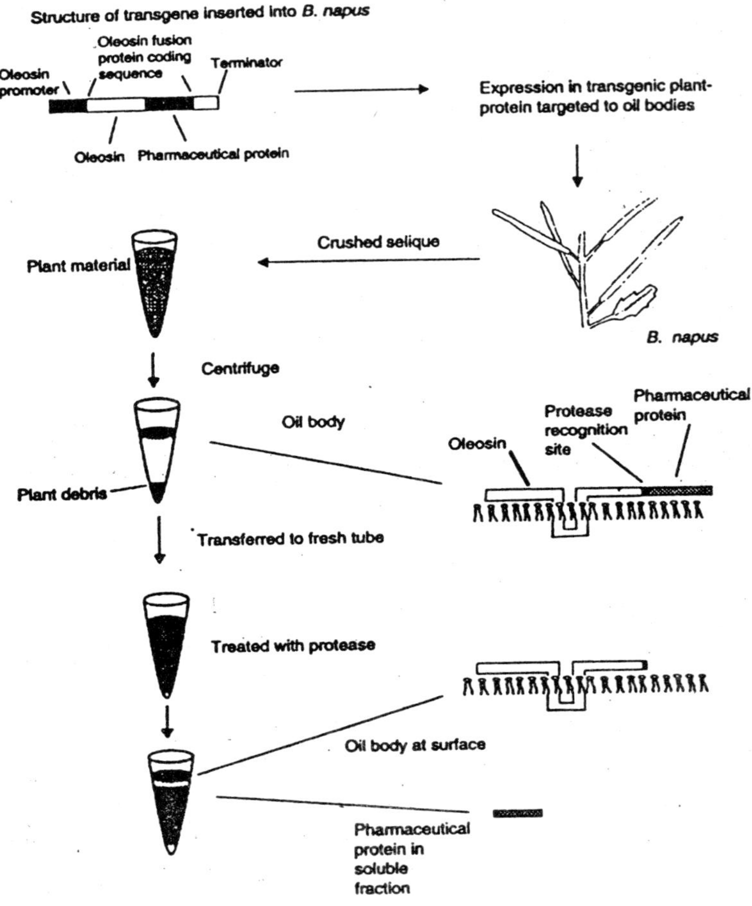

Figure 9.10: *A scheme for pharmaceutical protein purification by production in oil bodies. The pharmaceutical protein produced as a C-terminus fusion to oleosin (an oil body associated protein). The pharmaceutical protein and the oleosin are separated by a protease cleavage site. Seeds are crushed in an aqueous buffer and centrifuged, resulted in three fractions; the oil bodis, with the oleosins and the pharmaceutical protein, are the topmost layer (which floats on the aqueous layer). This oil body layer can be removed, re-centrifugee and then treated with a protease, which cleaves the pharmaceutical protein from the oleosin. Re-centrifugation results in highly purified protein being present in the aqueous phase. Full details can be found in the text. Redrawn from Goddijin OJM and Pen J. 1995 TIBTECH 13 379-387.*

target gene expression and protein production of specific tissues (see . Restricting gene expression to particular organs or tissues often leads to higher yields of recombinant proteins. Targeting expression has a number of advantages, some of which are related to the increase yield. The resources of the plant are effectively wasted if proteins are produced in parts of the plant that are not harvested. So, for vaccines that are to be ingested, after the minimum amount of processing, it makes sense to target the protein only to the parts of the plant that are normally eaten. Another important point is that storage organs, such as tubers and seeds, are designed to maintain their biological integrity over long periods. Targeting proteins to these structures has been shown to increase the stability of recombinant proteins from weeks to years. Seeds are also of great interest because they can offer simple strategies for protein purification. In tobacco, seeds contain a simpler mixture of proteins and lipids, with fewer phenolic compounds, that do green leaves.

As well as targeting expression to particular organs or tissues, intracellular targeting has been used to optimise the production of active proteins. Post-translational modification has been shown to be associated with the ER and/or apoplastic export, and the inclusion of transit peptides has been shown to improve both yield and activity. Tissues and intracellular targeting are brought together in the use of oleosins as vehicles for protein purification.

Transient expression

Plant virus capsids have also been used as carriers of recombinant proteins, particularly vaccines. In one approach, coding sequence for epitopes or proteins have been introduced into the coat protein gene of the virus genome so that it is 'presented' externally when virus particles are made. This approach is illustrated in. Another approach that has been used is to construct viral vectors to produce recombinant proteins that are targeted to the ER for processing. The virus can be replicated in the host plant, and through serial passage enough protein can be generated for clinical use.

The production of enzymes for industrial uses

The global industrial enzyme market has been estimated to be worth something approaching $US2 billion per annum. Proof of concept incursions by plant biotechnology into this market were on a small scale. The first commercialize 'industrial proteins' produced from transgenic plants summarises recombinant

TABLE 9.4 : EXAMPLES OF ENZYMES PRODUCED IN PLANTS

Enzyme	Use
Avidin	Diagnostic kits
β-Glucuronidase	Diagnostic kits
Trypsin	Pharmaceutical, wound care
Cellulase	Ethanol production from cellulose waste
Thermostable xylanase	Biomass processing
Phytase	Phytate breakdown, improved phosphate utilization
α-Amylase	Food processing
(1-3) (1-4) β-Glucanase	Brewing
Lignin peroxidase	Paper manufacture

enzyme production in plants) were two proteins familiar to the molecular biologist—avidin (from chicken) and β-glucuronidase (from *E.coli*)—both of which were produced in maize. Following on from these successes, ProdiGene Inc., the company responsible of plant-derived transgenic enzymes for commercial use. They announced in Spring 2002 that they would be going into large-scale production of trypsin, a proteolytic enzyme, which by its very nature is difficult to produce in conventional recombinant systmes. This protein is currently harvested from bovine and porcine pancreases, and is used for a number of applications, including the production of pharmaceutical such as insulin, vaccine production and wound care. The company has predicted that the worldwide demand for the enzyme will increase fivefold in the next 5 years. Existing sources may be unable to supply this increase, and there is the added worry of the spread of pathogen (like BSE) with products derived from animals. ProdiGene Inc.

aims to meet this demand by producing kilogram quantities of the protein from plant sources.

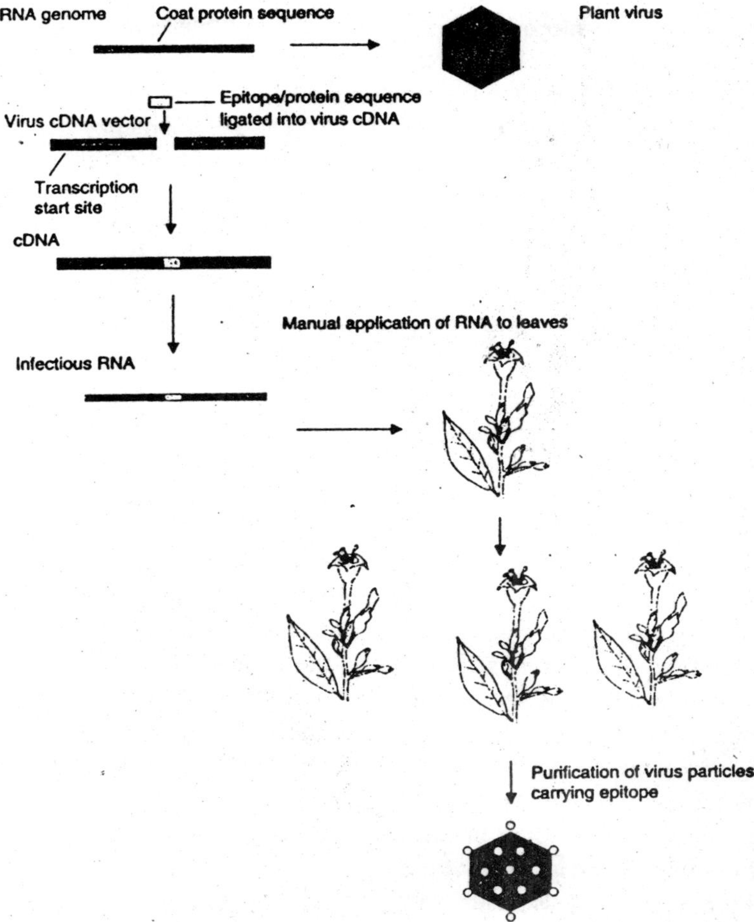

Figure 9.11: ***Construction of a recombinant plant virus carrying a protein epitope. The gene for the epitope was intoduce into the coat protein sequence within the vector. Infectious RNA was made and then introduced into plants (transfection) where virus capsids, displaying the expressed protein, were produced. This is known as the 'overcoat system' and can be used for a range of proteins.***

Some of the other important enzymes that may be produced in plants are, interestingly, normally targeted against components of the plant cell. Cellulases and xylanases,which are normally produced by the microorganisms associated with digestion in ruminants, have been produced in a number of different plants. There is a large market for these enzymes in the bioethanol, textile, pulp and paper industries and for the production of animal feed. If may, at first sight, seem a strange concept to express such enzymes in plants, due to the risk of 'autodigestion', but engineered, thermostable, forms of the enzymes, with high temperature optima were used. Therefore, no deleterious effects were detected at the temperatures at which a plant normally grows, but high levels of enzyme activity were detected when plant extracts were heated to the temperature optima of the enzyme. These were only small-scale initial trials, but the process could easily be scaled up. An interesting point to note is that these enzymes would require only the minimum of purification (an important consideration when determining applicability and economic viability). This is also the case for another important hydrolytic enzyme, phytase. This enzyme catalyses the hydrolysis of inositol and inorganic phosphate . Phytase is a very useful enzyme as it releases phosphate from a substrate that is normally indigestible to monogastric animal. As phytate is found in significant quantities in seeds used to formulated feed meal for pigs and poultry, these animals do not benefit from this source of phosphate, which is required for proper skeletal growth. Problems with phytate are exacerbated by the fact that monogastric animals excrete the indigestible phytate in their faeces, which contributes to an excessive phosphate build-up in ground water, leading to eutrophication. Transgenic plants expressing phytase in, for example, seeds have been produced. When these transgenic seeds are added to the diet of pigs and poultry, phosphate utilization is increased (negating the need for supplements) and phosphate excretion is decreased. Thus, the addition of the enzyme is a simple plant-derived formulation solved both a nutritional and an environmental problem.

Medically Related Proteins

This is an important, expanding, area for plant biotechnology, in which there is great hope that recombinant proteins made *in planta* will be accepted and widely utilized. This section of the chapter is divided into three sections:

(1) antibodies;

(2) vaccines; and

(3) other medically related proteins.

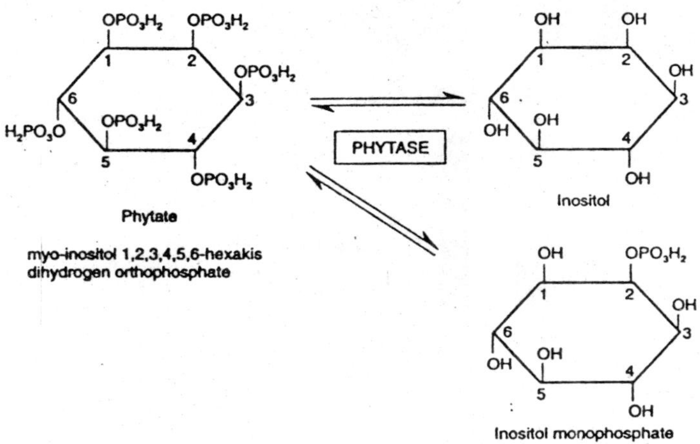

Figure 9.12: ***The action of phytase. Phytate is broken down either to inositol monophosphate (which is believed to be the main product of phytate breakdown) or to inositol.***

Antibodies or 'plantibodies'

In mammals there are a number of different types of antibodies, or immunoglobulins (lg). The immunoglobulins have been divided into five major classes on the basis of their physical, chemical, and immunological properties: IgA, IgD, IgE, IgG and IgM. The structure of IgG which can be used as an example. Immunoglobulins are composed of two heavy chains and two light chains. Each class of Ig has itw own class of H (heavy) chain, termed α, δ, ε, γ and μ, respectively. (IgA molecules have two α-class heavy chains, for example). Each Ig also has two light (L) chains, which are either κ or λ, but not a mix of the two. IgA, the main antibody found in secretions, can have a more complicated structure. In secretions,. IgA exists as a dimer of two IgA molecules joined by a J (or joining) chain and associated with a secretory

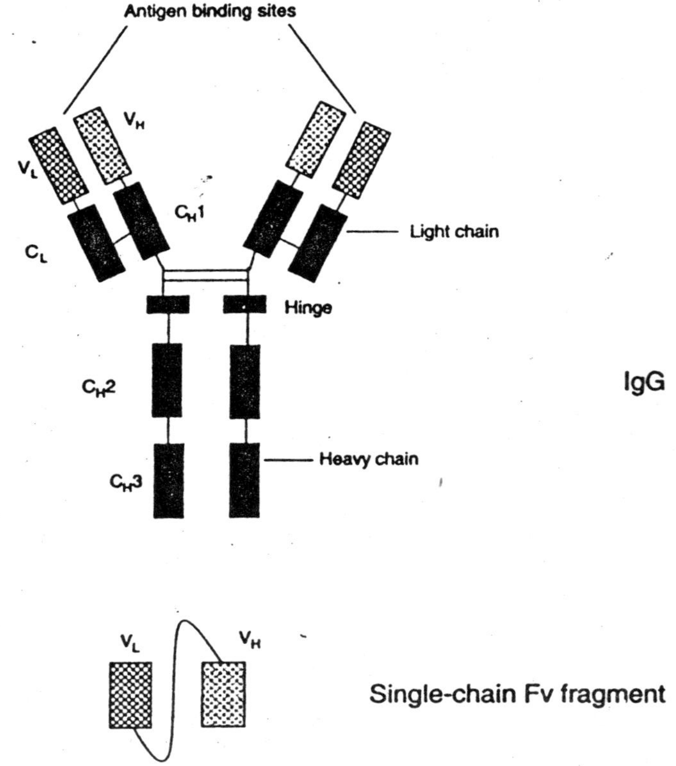

Figrure 9.13: ***Structures of IgG and an engineered single-chain varibale fragment antibody. A typical antibody is composed of two identical heavy (H) chains and two identical light (L) chains. Antigen-bindings sites are found at the amino-terminal regions of both the H and L chains. The hinge and tail regions are formed only from the H chains. Thelight and heavy chains are composed of constant (C) and variable (V) domains. The variable domains of the heavy (V_H) and light (V_L) chains make up the antigen-binding sites. Single-chain variable fragment (scFv) antibodies are engineered by the joining together light and heavy chain variable regions. Antigen-binding is determined solely by these variable regions.***

component. Each antibody has two antigen-binding sites, formed from the varibale (V) regions of the light and heavy chains. As the variable regions of the light and heavy chains are solely responsible for antigen binding, simpler, smaller molecues that still bind antigens can be produced. Single-chain variable fragment antibodies are produced from synthetic genes made by fusing the sequence for light- and heavy-chain variable regions. Although not the only example of engineered antibodies, scFv antibodies are the most commonly used and most successful.

The first example of a functional antibody (a mouse immunoglobulin IgG1) being produced in plants are reported in 1989. The construction of this antibody was a two-step process. In the first stage, two separate transgenic tobacco lines were produced, one contained the gene for the γ-heavy chain the other contained the gene for the κ-light chain. The second stage of the process was to cross these plants to generate progeny that expresses both genes. It was found that in the F1 progeny expressing antibody were significant amounts (up to 1.3% of total leaf protein) of assembled antibody were produced. These transgenes contained native immunoglobulin signal sequences, which targeted the nascent light and heavy chains to the ER and were found to be necessary for efficient assembly of the antibody in plants.

Several important observations, which underpinned much of the future work in this area, were made in this original research. Targeting of the antibody chains to the lumen of the ER was found to be necessary for efficient chain assembly and stability. Conditions in the lumen of the ER favour the correct processing of the antibody chains, especially disulphide bridge formation, and molecular chaperones also help to direct antibody chain folding and assembly. Another important observation was that the formation of the complete antibody in the F1 plants increased the protein yield over that of either of the parents that contained singly chains, reflecting the stable nature of the complete antibody in the plant system.

Although a pioneering study, demonstrating that funcitonal antibodies could be assembled in plants, the antibody in question has no practical applications.

Antibodies produced in plants are thought to be particularly suitable for topical immunotherapy. In an extension of the strategy of producing the various antibody subunits in different plant lines that are subsequently crossed to produce a functional antibody, a secretory IgA (sIgA) that protects against dental caries produced by *Streptococcus mutans* has been expressed in plants. The antibody does this by recognizing the native streptococcal antigen (SA) I/II cell-surface adhesion molecule, thereby preventing colonisation.

TABLE 9.5 : EXAMPLES OF ANTIBODIES THAT HAVE BEEN PRODUCED IN PLANS

Application	Plant	Antibody	Signal sequence
Immunoglobulins			
S. mutans SA I/II (dental caries)	Tobacco	sIgA (hybrid)	Murine IgG
S. mutans SA I/II (dental caries)	Tobacco	IgG (guy's 13)	Murine IgG
Surface antigen (colon cancer)	Tobacco	IgG Co 17-1 A	Murine IgG/KDEL
Herpes simplex virus	Soybean	IgG (anti HSV-2)	Tobacco extensin
Single-chain Fv			
Lymphoma	Tobacco	scFv (38C13)	Rice a*-amylase
Carcinoembryonic antigen (cancer)	Cereals	scFvT84.66	Murine IgG/KDEL

This multiple-components IgA antibody required four parental transgenic tobacco lines to be produced, in which cDNAs for either the H, L or J chains (the H and L chains were derived from a murine antibody termed 'Guy's 13' that recognizes the SA I/II cell-surface adhesion molecule), or the secretory component, were expressed under control of the 35S promoter, and then crossed. The strategy used is illustrated in. The success of the approach has been quite an impressive technical achievement. In mammalian systems, both plasma cells and epithelial cells are involved in the synthesis and assembly of secretory antibodies, yet single plant cell containing all four transgenes are able to produce them efficiently. Secretory antibodies have many theoretical advantages in situations like the one described here. Secretory IgAs are the predominant antibody type that protects

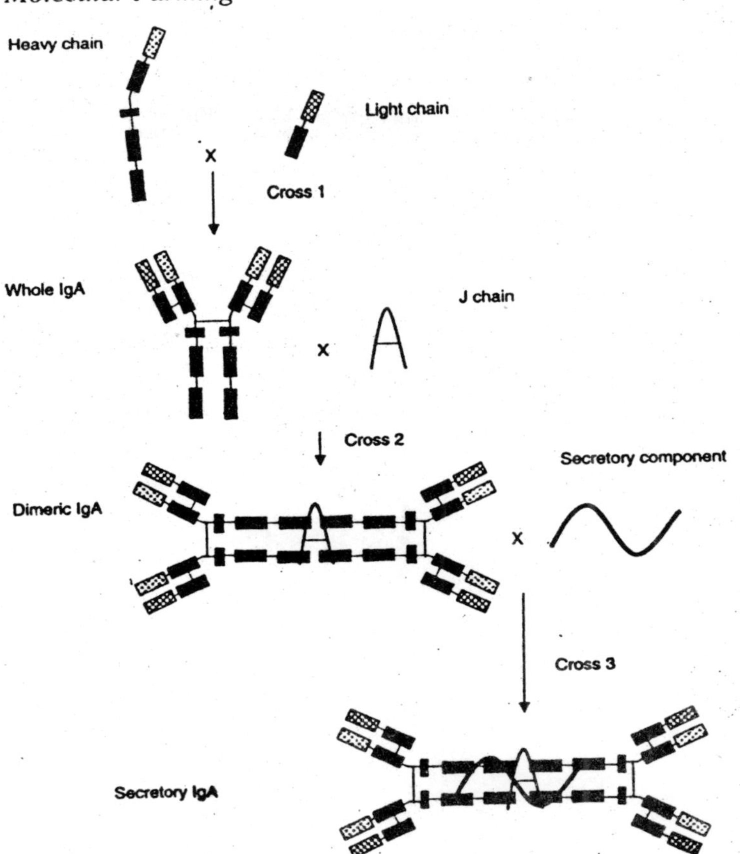

Figure 9.14: ***Synthesis of a secretory IgA (sIgA) molecules in transgenic plants. Four separate parental plants lines were constructed. Each expressed a separate piece of the antibody (the H, L or J chain, or the secretory component). The lines expressing the H and L chains were crossed, giving F1 plants that accumulated IgA. These plants were further crossed with plants expressing the J (joining) chain, which is, in progeny, resulted in the accumulation of dimeric IgA. These plants were finally crossed with plans expressing the secretory component. Progeny of this cross accumulated functional sIgA.***

against microbial infections at mucosal sites. They are also more resistant to proteolysis (which contributes to high yield), and they bind antigens with a higher efficiency than do the original antibodies produced in plants.

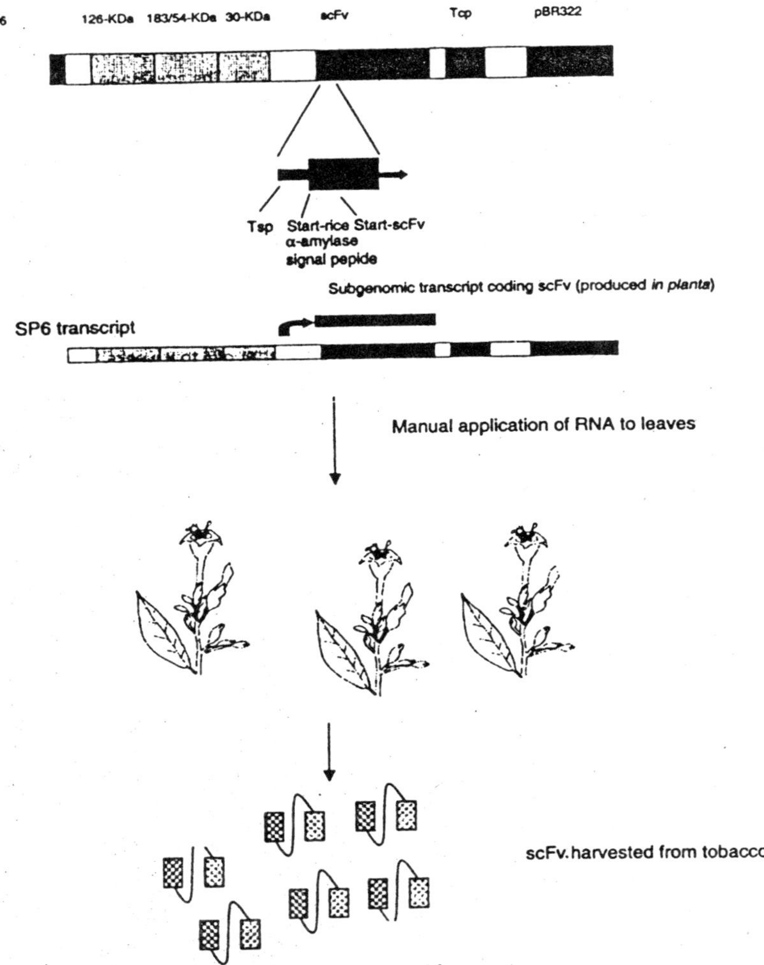

Figure 9.15: *The use of engineered tobacco mosaic virus (TMV) vectors for the expression of pharmaceutical proteins. An scFv antibody sequency has been located adjacent to a transplation start sequence and a rice a*-amylase signal peptide. Infectious RNA is produced from an SP6 promoter and used to infect plants (transfection). Subgenomic RNA is produced, and this acts as a template for expression of the scFv. scFv. proteins can be isolated. (Tsp, translation start position; Tcp, tomato coat protein; pBR322, bacterial replication sequence).*

These secretory antibodies have now been tested on humans. They have been topically applied to teeth and found to be effective at preventing recolonisation by *S.mutans* for up to 4 months, which is as effective as an IgG produced in a murine hybridoma. Despite the differences in structure, there appear to be no differences in binding properties between the proteins.

Plants have been used to produce a variety of antibodies, including whole antibodies, antigen-binding fragments and single-chain variable fragment antibodies. Despite the now well-characterized differences in the glycosylation pattern seen between antibodies produced in plant and mammalian expression systems, antibodies produced in plants generally seem to exhibit similar properties to antibodies produced in other, mammalian, systems. Some of the most important examples are lested in Table 9.5.

Plant-derived vaccines

Both stable and transient expression systems have been used to produce vaccines in plants. One of the first successful attempts in this area used a transient expression system. Recombinant cowpea mosaic *comovirus* (CPMV) was used as a vector for a linear epitope (antigenic determinant) from the VP2 capsid protein of mink enteritis virus (MEV). The recombinant CPMV was replicated in black-eyed beans (*Vigna unguiculata*), from which chimeric virus particles (CPMV-VP2), displaying the VP2 epitope, were isolated and used for the subcutaneous injection of mink. The vaccine was able to induce resistance to clinical infection by MEV. Many other vaccines have now been produced in plants and Table 9.6 lists many of these.

The main aim of many of these studies is the generation of edible vaccines. The driving force behind the use of edible vaccines has been the need for cheap, effective treatments for enteric disease in the developing world. Plant-based systems have the potential to provide low-cost, easily administered, locally produced edible vaccines that could provide mucosal immunity against the infectious agents that are responsible for the death of millions of people, particularly children, annually. To bring this about it is necessary to present, in a practical, edible form, a quantity of vaccine that will stimulate the GALT (gut-associated lymphoid tissue) system in the gut to generate antibodies. The question that have been had to be considered relate to delivery of the vaccine to the gut through the stomach. As proteins, these vaccines are likely to be degraded in the stomach. The experiments carried out so far indicate that orally administered plant material expressing a vaccine can induce an immune response, with particulate material the most effective. Interestingly, when virus coat proteins are expressed in plant leaves they can form virus particles that may also improve the stability and effectiveness of the vaccine by presenting the protein in its native form.

Much of the initial work on producing vaccines in plants was carried out in inedible plants, such as tobacco. There has been a move by some scientists towards producing vaccines in plants that could be eaten raw, such as tomato and banana. Bananas are considered to be the ideal production systems by some scientists as they are grown widely in the countries of the developing world and can be eaten raw. For use in animals, it is possible to consider the use of fodder crops or other food crops.

These so-called 'edible vaccines' do work in the laboratory, but would they really work as originally envisaged? The major flaw in the process is that one needs to control very carefully the dose being given to patients if effective and safe protection against disease is to be ensured. One of the drawbacks of edible vaccines is that it may prove very difficult to control the dose administered if consumption as part of a foodstuff is the means of delivery. Charles Arntzen, one of the main lights in this field, has recently withdrawn from the concept of edible vaccines. He believes they should be considered vaccines that come from a plant source. His group produces vaccines in tomato, but they are partially processed so that food-based tablets containing a known dose of vaccine can be produced. Not only is this better science but it will be easier to license the use of these vaccines from plant sources. Bananas may, of course, still prove to be a useful system that allows locally based production facilities to be established in developing countries.

TABLE 9.6 : EXAMPLES OF VACCINES PRODUCED IN PLANTS

Origin	*Recombinant protein*	*Plant*	*Production level*
Human			
Escherichia coli	Heat-labile enterotoxin B	Tobacco	0.001% SLP
Vibrio cholerae	Cholera CtoxA and CtoxB subunit	Potato	0.3% TSP
Hepatitis B	Envelope surface protein	Tobacco/potato	<0.1% FW
Norwalk virus	Capsid protein	Tobacco/potato	0.23%/0.37% TSP
Rabies virus	Rabies virus glycoprotein	Tomato	1% TSP
Animal			
Foot and mouth virus	Virus epitope VP1	Alfalfa/*Arabidopsis*	N/A
Porcine coronavirus	Viral glycoprotein	Tobacco/maize	0.2% TSP/0.01% FW
Mink enteritis virus	Viral epitope VP2	Blackeyed bean	N/A (CPMV)
Canine parvovirus	Peptide from VP2 capsid protein	*Arabidopsis*	3% SLP

Other proteins—biopharmaceuticals

A wide range of proteins of pharmaceutical interest has been expressed in transgenic plants in the hope of producing an economically viable system for large-scale prodution. Although active recombinant proteins have been produced, one problem associated with production in plant systems is that these often give a relatively low product yield and recovery. Various strategies to overcome this problem are being developed, amongst them the use of novel purification systems (see the section on oleosins for an example) and chloroplast transformation.

Despite some difficulties, plants hold great promise as production systems for pharmaceutical proteins, and we will look at some examples in a little more detail.

Trichosanthin

Trichosanthin (TCS) is a component of the tuber of the plant *Trichosanthes kirilowii*, which is used in Chinese medicine. A ribosome-inactivating protein, TCS inhibits tumour growth and the immune response, and may be useful as a treatment for HIV/AIDS. Relatively high level of TCS accumulation (2% of total soluble protein 2 weeks after inoculation) have been achieved in *Nicotiana benthamiana* using a viral RNA-based transfection system. This system offers the possibility of rapid, large-scale production of TCS. The rapid production also offers the opportunity to rapidly produce mutated forms of TCS for screening in trials aimed at identifying forms with increased efficacy or reduced side-effects.

Glucocerebrosidase

Gaucher's disease is an inherited disorder in which glucocerebroside (a component of red blood cells that is normally broken down into glucose and ceramide as old and damaged red blood cells are removed by the body) accumulated in lysosomes, due to a deficiency in the enzyme glucocerebroside. Gaucher's disease, which results in swelling of the spleen and liver and severe bone damage, can be extremely debilitating and painful. Currently, treatment is based on managing the symptoms or by treatment with a drug developed from glucocerebroside purified from human placentas. However, a huge amount is required (10-12 tons of placentas a year for one patient), making this one of the world's most expensive drugs. Recombinant approaches to production in mammalian cell cultures have reduced the cost to some extent, although it remains an extremely expensive drug.

TABLE 9.7 : EXAMPLES OF BIOPHARAMECUTICALS PRODUCED IN PLANTS

Recombinant protein	Origin	Plant	Application
Protein C	Human	Tobacco	Anticoagulant
Hirudin	*Hirudo* medicinalis	Canola	Anticoagulant

Somatotrophin	Human	Tobacco	Growth hormone
β-Inteferon	Human	Rice/turnip/tobacco	Treatment for hepatitis B + C
Serum albumin	Human	Tobacco	Burns/fluid replacement, etc.
Haemoglobin-α and β	Human	Tobacco	Blood substitute
Homotrimeric collagen	Human	Tobacco	Collagen
α_1-Antitrypsin	Human	Rice	Cystic fibrosis, haemorrhages
Aprotinin (trypsin inhibitor)	Human	Maize	Transplant surgery
Lactoferrin	Human	Potato	Antimicrobial
ACE	Human	Tobacco/tomato	Hypertension
Enkephalin	Human	*Arabidopsis*/canola	Opiate
Trichosanthin-α	*Trichosanthes kirilowii*	Tobacco	HIV therapy, cancer

A process to produce glucocerebroside in tobacco has recently been patented, which will hopefully result in cheaper glucocerebroside being available for patients with Gaucher's disease.

Human serum albumin

Human serum albumin (HSA) has many potential medical applications (the treatment of burns and in liver cirrhosis, for example). HSA has been expressed in tobacco and potato under the control of the 35S promoter. Two forms of the HSA protein were expressed, with different signal sequences to ensure secretion of the HSA. One form had the human prepro-sequence, whilst the other had the signal sequence from the tobacco extracellular PR-S protein. Although both forms of the HSA were secreted, analysis showed that the form with the human prepro-sequence was not properly processed, leading to proHSA being secreted. However, the form with the PR-S signal sequence was correctly processed, and mature HAS, that was indistinguishable from the human protein, was produced.

Arginine vasopressin

The examples we have looked at demonstrate that both plant and non-plant pharmaceutical proteins, which are of high value both clinically and economically, can be successfully produced in plants. However, in what is a typical situation in developing technologies, not all attempts at protein production in plants have been successful.

Attempts to express arginine vasopressin (AVP) in tobacco cultures, for example, have proved unsuccessful due to protein instability and retarded cell growth.

ECONOMIC CONSIDERATIONS

Whilst there has been much talk in this chapter of suitable targets for molecular farming, few of the examples described have yet been produced commercially. This may result from problems with the technology itself(see the section on cyclodextrins), or it could reflect the long time between initial laboratory-base demonstration and subsequent optimisation and regulatory approvals for field release and approval for food or drug use. However, it may be the case that levels of production that represent major scientific breakthrough are none the less insufficient to be economically viable. This may seem surprising, because it is tempting to make the naive assumption that the production of compounds in a field, with free energy from the sun, must be cheaper than production in a chemical factory or industrial fermenter. This is often not the case, because farming is not a cost-free activity, the extraction of the product from the plants may be a difficult and expensive process and the price of the product must be significantly higher than that of the crops it has replaced. This may mean that the current levels of synthesis in transgenic plants coupled to the costs of production make a number of the examples described above uneconomic when compared to current production methods. On the other hand, continuing improvements in plant biotechnology, coupled to the inevitable rise in costs of non-renewable resources, such as petrochemicals, make it likely that many of these products will eventually be made in plants.

The economics of producing medically related proteins in plants has been much discussed, particularly becasue of the high unit price of many of these proteins and the expansion of the market for medically related proteins. It is in this area that many companies and people have invested money, in the hope of achieving economically viable molecular farming in plants. Fortunately for these companies and investors, the figures seem to suggest that there is a future for molecular farming. It has been estimated that the cost of producing immunoglobulins in alfalfa, in a 250-m^2 greenhouse, are %US500-600g^{-1} compared with figure of \$US5000g^{-1} for a hybridoma-produced antibody. The company Planet Biotechnology has compared the costs of producing and IgA antibody in four different systems. The comparison was between production in a cell culture system, transgenic goats, grain (assuming a yield of 7.5 tonnes ha^{-1}) and green biomass, such as leaves (assuming a yield of 120 tonnes ha^{-1}). The cost of an IgA produced in leaves was estimated to be below \$US50 g^{-1}. This compared very favourably with the costs for production in a cell culture system (\$US1000g^{-1}) and in a transgenic animal (\$US100 g^{-1}). It must be remembered that these figures are fairly crude, being based on projected yields, etc. However, they do illustrate that costs of protein production in plants are at least comparable with other systems of production and further developments are now coming on stream. Transformation systems are now available for the introduction of foreign genes into chloroplast, giving an estimated 10,000 copies per cell. Using such systems it is claimed that recombinant protein content has reached as much as 47% of the total soluble protein. An example given in. Somatotrophin (-growth hormone), produce in chloroplast, reached about 7% of the TSP. This compares favourable with that of a nuclear transgene (0.01% TSP). Using Chloroplast based plant systems have other potential advantage over existing systems. Despite the very high number of gene copies there have been no reports of gene silencing that could effect expression levels. Also, protein processing is one of the most expensive parts of producing proteins in existing systems such as *E.coli*. It seems that chloroplast can assemble and fold complex foreign proteins, but whether they can copy with the other post translational modification problems remain to be seen. Another advantage of chloroplast is that they are meternally inherited. This means that they will not be spread via pollen to non-transgenic systems. This type of clean gene technology is most important for getting approval for growth in non-contained facilites. Other Health related proteins produced in chloroplasts include: human proinsulin for the treatment of autoimmune disease, Interferon IFNa 2b, human serum albumen, synthetic human haemoglobin genes, Guys 13 monoclonal antibody, and a *Bacillus anthracis* protective antigen.

10

Genetic Improvement of Crops

For thousands of years agricultural people slowly improved their crops, either unwittingly of sometimes quite consciously, by setting aside part of the harvest as the planting material for the next crop. Crop improvement was greatly accelerated in the 20th century by the rediscovery of the laws of inheritance, first formulated by Gregor Mendel in 1865, and after scientists established that the chromosomes contain the information for specific traits encoded in DNA. The rise of molecular biology, starting in the 1950s with the discovery of DNA's structure, led to a series of further discoveries showing how traits are encoded as proteins in the cell and the plant. This understanding, and new techniques that allow scientists to manipulate DNA in the laboratory, are revolutionizing plant breeding.

In the future, increasing yield for traditional crops, improving nutritional or postharvest quality of these crops, adapting them to more stressful environment, domesticating new crops, and converting existing ones to plant factories that produce chemicals for industry with all flow from applying the principles of plant breeding, genetics, molecular biology, and genomics. "Classic" plant breeders—those who perform mostly field work—will only succeed in making major improvements by exploiting more and more of the tools used in molecular biology. Collaboration between plant breeders, plant pathologists, molecular biologists and geneticist will be essential to expedite the long process of producing new and improved varieties.

GENES ARE MADE OF DNA

The 30-year period from the 1930s to the 1960s was a very exciting time for scientific discovery, and must go down in history when one of the most fundamental questions science ever asked was answered. What exactly are genes made of?

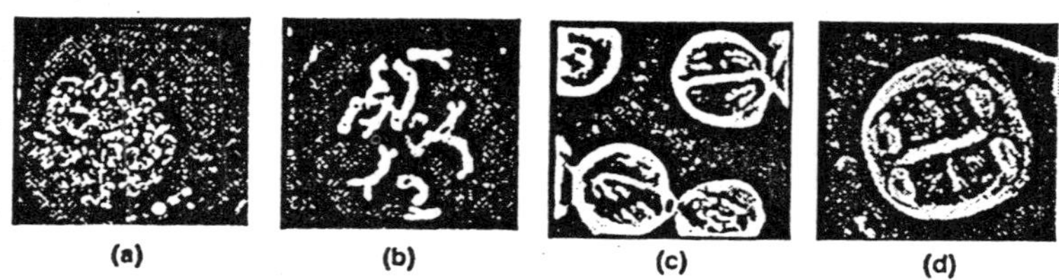

Figure 10.1: ***Structure of chromosomes in microsporocytes of lily. Microsporocytes are the precursorts of the sex or gametes that are formed in flowers anther. The cells shown here have been treated with a dye that binds to DNA and allows all DNA-containing structures to be seen clearly. In (a) the chromosomes have just started tocontract and are visible as long thin threads. In (b) contraction is at its maxium and each chromosome is clearly distinguishable. In (c) thechromosomes are separating from each other during the second meiotic division, and in (d) theformation of the four gametes is nearly complete.***

Careful observation of how chromosomes separate during mitosis, and the discovery that sex cells contain only half the number of chromosomes (one set instead ot two) that other cells have, led scientists to postulate that Mendel's inherited units, later called *genes*, are organized on the chromosomes. Chromosomes are the filamentous structures in nuclei. Geneticists knew that chromosomes were composed of both protein and DNA (deoxyribonuleic acid). For many years, biochemists incorrectly believed that genes consisted of protein because they knew proteins were large, complex molecules (they consist of countless sequences of 20 different amino acids) that show great specificity in their activity (mostly enzymatic activity). Biochemist thought that only protein molecules could contain enough information to specify inherited traits.

Adenine

Guanine

Deoxyadenosine monophosphate

Deoxyguanosine monophosphate

(a) Purine nucleotides

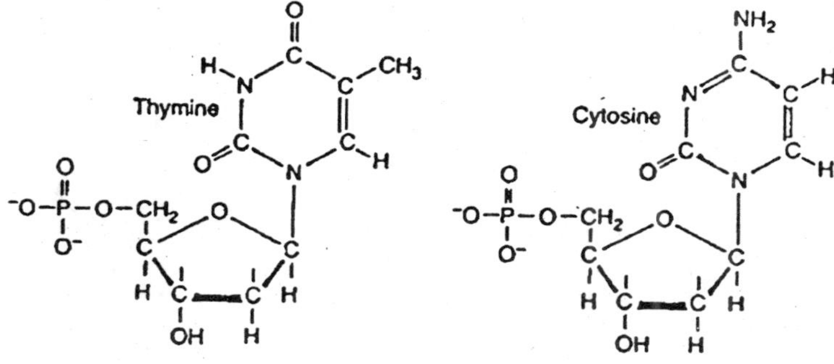

Figure 10.2: ***The building blocks of DNA. DNA molecules are polymers, built out of monomers called nucleotides. (a) The double-ring purine nucleotides and (b) the single-ring pyrimidine nucleotides each have a phosphate, the five-carbon sugar, deoxyribose, and one of four nitrogen-containing bases.***

In contrast, they thought DNA was a small and relatively simple molecule because it contains only four nucleotides—surely DNA could not contain enough information to account for the great diversity of life forms on Earth. Several elegant and decisive experiments conducted with bacteria and viruses by Frederick Griffin, Olwald Avery, Alfred Hershey, and Martha Chase showed that genes consist of DNA and have the necessary information to direct the synthesis of proteins. The researchers showed that mutation of bacterial genes results in the loss of specific enzymatic activities, indicating that genes are responsible for the presence of proteins in the cell. The discovery later led to the "one gene, one enzyme" hypothesis that a distinct gene controls the presence of each enzyme in the cell.

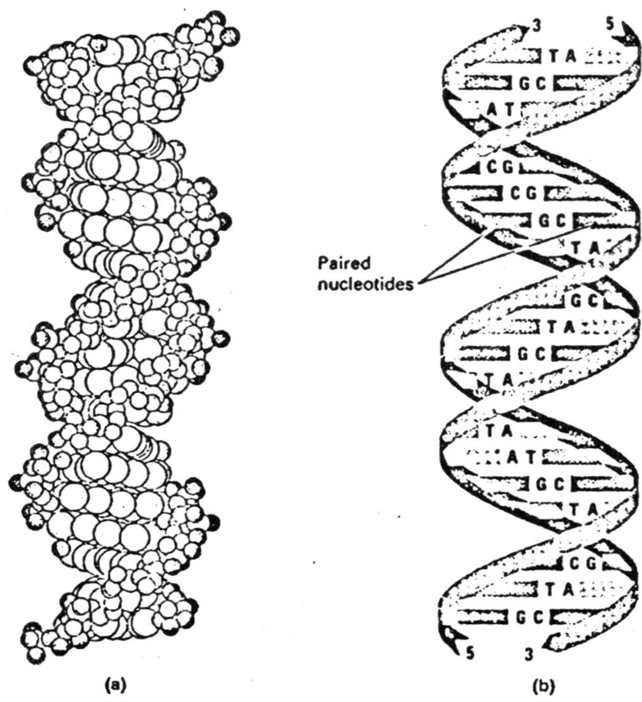

Figure 10.3: ***The structure of DNA. (a) A space-filling model in which each atom appears as a sphere. DNA molecules are polymers made by linking the phosphate on one nucleotide to the oxygen on a specific carbon of the deoxyribose of the next nucleotide in the chain. This produces a molecule that consists of a backbone of alternating phosphate and deoxyribose groups, with the bases protruding. Molecules of RNA are built in a similar fashion, but in a complete DNA molecule (as shown here), two nucleotide chains pair to form a doubled helix. (b) The diagram shows a segment of double-stranded DNA that contains only 18 paired nucleotides; however, DNA molecules normally can contain thousands. A small gene may contain 2,500 nucleotides; a large one, 25,000 or more. The deoxyribose-phosphate backbone of each chain is on the outside, and the bases are on the inside of the double helix. The bases of one chain are paired with those of the other chain according to the base-pairing ruls: thymines aloways pair with adenines and cytosines with guanines.***

Now biologists were convinced DNA, not protein, was the genetic material, biochemist wanted to answer another fundamental question: What precisely is the structure of DNA? Because they knew this discovery would go down in history, researchers were racing each other to unravel the structure of DNA.

With new chemistries developed in the 1940s, Erwin Chargaff determined in detail the base composition of DNA. Scientists already knew that DNA consists of four different types of nucleotides: two with the purine bases adenine and guanine, and two with the pyrimidine bases thymine and cytosine. The purine bases have a double ring structure, whereas the pyrimidine bases have a singly ring. These four nucleotides—the building blocks of DNA—consist of a phosphate group, the sugar deoxyribose, and one of the four nitrogen-containing bases (A, C, G, or T). Chargaff showed that in some species, such as *E.coli* and maize, the DNA consists of about 25% of each type of nucleotide, but that in most species, such as humans, this is not the case. Thus DNA shows the variability among species that you might expect of the genetic material life. Within a species, however, Chargaff's experiments showed that the amount of A always equals the amount of T, the amount of G always equal the amount of C, and that the percentage of A plus G equals to 50% and the percentage of C plus T equals 50%. These data suggested that A is always paired with T and that G is always paired with C.

Rosalind Franklin studied the structure of DNA using X-rays and found that DNA has a helical structure. Using Chargaff's and Franklin's data,and possible some of Linus Pauling's data, James Watson and Francis Crick constructed a model for the structure of DNA, for which they won the Nobel Prize in medicine in 1962. They concluded that DNA is a double-stranded helix with the adenine in one strand always hydrogen-bonded to the thymine in another strand; similarly, cytosine in one strand is always hydrogen-bonded to guanine in the other strand. Together the two strands form the double helix with the sugar-phosphate backbones on the outside and the paired bases on the inside. The two strands of the DNA molecule are said to be *complementary strands* because if you know the base sequence of one strand, you can deduce the

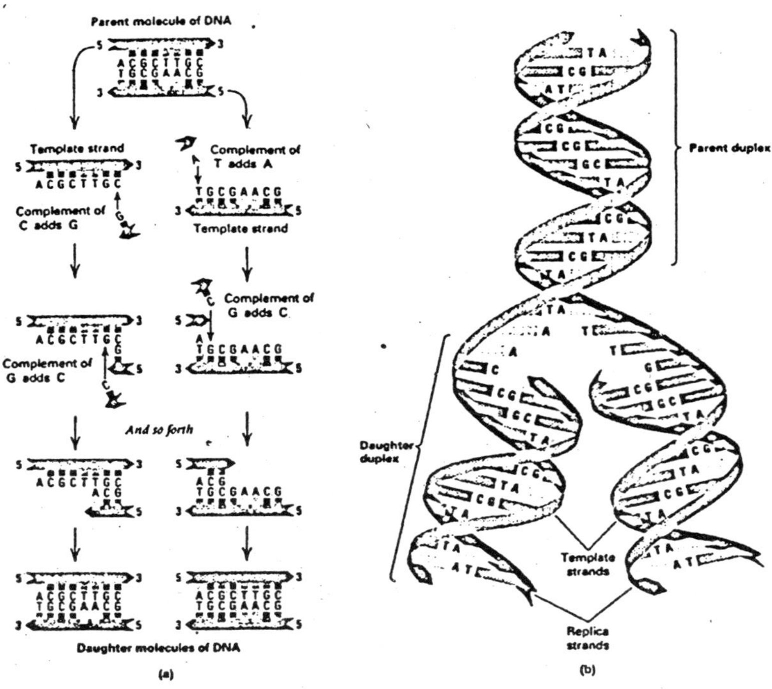

Figure 10.4: ***Replication of DNA. (a) When DNA replicates, the two strands of the molecule separate and each serves as a template for the synthesis of a new complementary strand, which grows in length by the successive addition of a single nucleotide. (b) The obligatory pairing of adenines with thymines and guanines with cytosines ensures that the two daughter molecules will be duplicates of the original.***

base sequence in the other complementary strand. If one strand contains the sequence ATTGCC, than the other strand must in the same region have the sequence TAACGG, because of the base-pairing rules (A opposite T, and G opposite C).

Another requirement of the genetic material is that it must replicate and then transfer to daughter cells. Watson and Crick proposed a mechanism for the copying of DNA based on the specific base-pairing rules, and scientists have now confirmed that DNA replicates by complementary base pairing. DNA copying is called *semiconservative replication* because one of the old strands is present in each new daughter molecule. When DNA replicates, the two old strands separate and the complementary strands then synthesize using the existing strands as templates. When the two new strands are synthesized, A is always opposite T and C opposite G, so that two identical molecules form. Because the information is contained in the exact sequence of the bases, DNA replication makes a new and exact copy of the information.

Most DNA in a plant cell is found in the nucleus and the totality of this DNA is called *nuclear genome*. Two cellular organelles, the chloroplasts and the mitochondria, have their own DNA—their own genomes—which encodes a number of genes. Mitochondrial DNA encodes about 15 genes, and chloroplast DNA, about 100 genes. Chloroplasts and mitochondria are the evolutionary descendents of endosymbiotic bacteria, and they still share many properties with present-day bacteria. Mitochondrial DNA and chloroplast DNA are inherited in a uniparental manner, usually with the female sex cell and not in Mendelian fashion (see following discussion). Transforming plants and creating genetically modified (GM) crops usually involves introducing DNA into the nuclear DNA of plants.. Recently it has also become possible to introduce new genes into chloroplast DNA

At one point it was thought that the nuclear genome of an organism is completely stable concept for the occasional mutations caused by chemicals or irradiation. Now we know that the chromosomes contain elements that move from place to place by the shuffling of these mobile elements.

Genes Code for Protein Via the Molecular Trilogy: DNA, RNA, and Amino Acids.

Proteins are linear molecules of 20 different amino acids, usually between 100 and 10,000 of them, folded in a three-dimensional configuration that is often globular in overall shape. The individual amino acids join together by linkages called *peptide bonds*, so a protein molecule is often called a *polypeptide*. This string of amino acids then winds into a helix or can take the shape of a pleated sheet. When the entire chain is folded, some parts will be helical, other parts will be pleated, and yet other portions of the chain will form a random coil.

H_2O

Dipeptide

(a) Peptide bond Peptide group

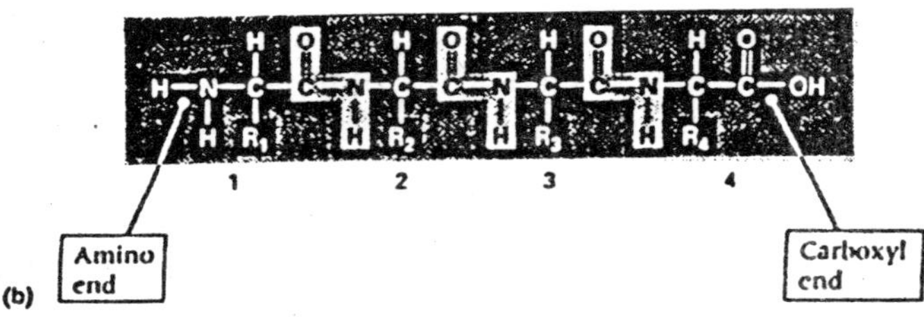

Figure 10.5: ***Properties of a polypeptide chain. (a) Formation of a dipeptide by reaction of the carboxyl group of one amino acid (top) with theamino group of a second amino acid (bottom). A water molecule is removed to form the peptide bond (red line). (b) A tetrapeptide showing the alternation of a*-carbon atoms (black) and peptide groups (blue). The four amino acids are specified by the R groups.***

Proteins may interact with other molecules in the cell in very specific ways: as enzymes, as regulators of gene activity, or as transporters of ions and water. This specificity resides in their exact three-dimensional shape, which is in turn determined by the sequence of amino acids. Thus each kind of protein (and some cells probably have more than 5,000 different proteins) has a characteristic amino acid sequence. The information that specifies the order in which the amino acids must assemble when a protein is synthesized, is contained in the DNA, specifically in the sequence of bases attacked to the sugar-phosphate backbone.

At first sight this coding might seem be to a problem, because DNA has only four different bases and proteins have 20 different amino acids. This coding problem is solved by having a sequence of three bases specify one amino acid. When four different nucleotides are arranged in groups of three, 64 different combinations (4 × 4 × 4) are possible, more than enough to specify 20 amino acids. In the 1960s, Marshall Nirenberg, Heinrich Matthei, and

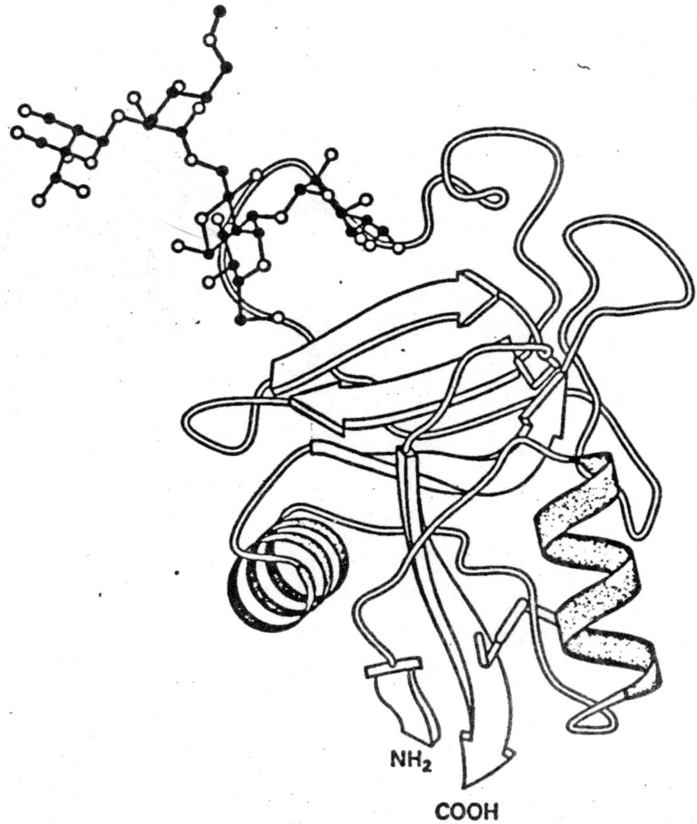

Figure 10.6: ***A ribbon diagram showing the ways in which a polypeptide can be folded. Arrows represent paralled b* sheets, and helical regions are shown as coiled ribbons. The protein in this example is a protein that binds to carbohydrates. The stick figure in the upper left is a small oligosaccharide consisting of several sugars bound to the protein.***

Philip Leader figured out which three-base combination or codon specifies a particular amino acid. The correspondence between a three-base combination and an amino acid is called the *genetic code*. The genetic code is universal: it is the same in all organisms. Most amino acids are represented by more than one codon. Of the 64 different prossibilities, 61 codons specify amino acids,and 3 specify "stop" signals marking the end of a protein-coding segment of the DNA.

TABLE 10.1 : THE GENTIC CODE

	Second Position				
First Position	**U**	**C**	**A**	**G**	**Third position**
	Phe	Ser	Tyr	Cys	U
	Phe	Ser	Tyr	Cys	C
U	Leu	Ser	Stop	Stop	A
	Leu	Ser	Stop	Trp	G
	Leu	Pro	His	Arg	U
	Leu	Pro	His	Arg	C
C	Leu	Pro	Gln	Arg	A
	Leu	Pro	Gln	Arg	G
	lle	Thr	Asn	Ser	U
	lle	Thr	Asn	Ser	C
A	lle	Thr	Lys	Arg	A
	Met	Thr	Lys	Arg	G
	Val	Ala	Asp	Gly	U
	Val	Ala	Asp	Gly	C
G	Val	Ala	Glu	Gly	A
	Val	Ala	Glu	Gly	G

Note: A sequence of three nucleotides forms the nucleic acid codon fo ra singly amino acid. The four nucleotids U, C, A ,G cna produe 64 different three-nucleotide combinatins. All the amino acids except methionine (Met) nad tryptophan (TrP) have more than one codon. The "stop" codons UAA, UAG, and UGA do not code for amino acids but signal the end of a protein. All proteins start with methionine.

The codons are given as they appear in messenger RNA. The four bases in the nucleotides of ribonuleic acids are uracil (U), cytosine (C), adenosine (A), and guanine (G). The amino acids specified by the genetic code are alanine (Ala), arginine (Arg), asporagine (Asn), aspartic acid (Asp), cysteine (Cys), glycine (Gly), glutamie (Gln), glutamic acid (Glu), histihine (His), isoleucine (lle), leucine (Leu), lysine (Lys), methionine (Met), phenylalanie (Phe), proline (Pro), serine (Ser), thronine (Thr), tryptophan (Trp), tyrosine (Tyr), and valilne (Val).

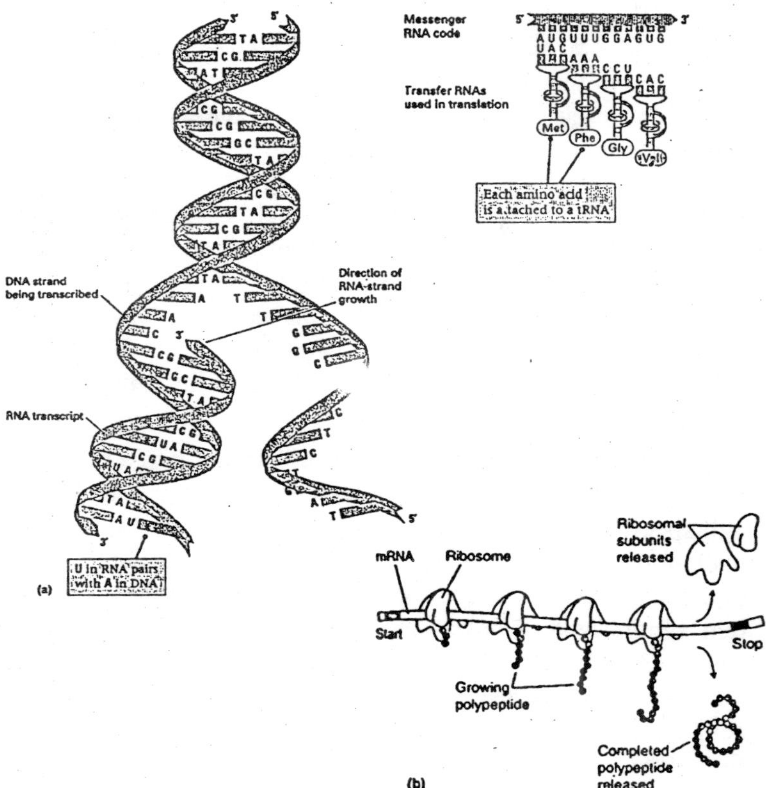

Figure 10.7: ***Protein synthesis. (a) Before protein synthesis can begin, the genetic information (sequence of bases) in the DNA must be copied into an RNA molecule. The two strands of the DNA molecule separate, allowing one strand to becopied into a messenger RNA (mRNA). This process, called transcription, takes place in the nucleus. The mRNA moves from the nucleus to the cytoplasm, where proein synthesis occurs. This coincides with a maturation process in which parts of the mRNA that do not code for protein are removed. In thecytoplasm, specific enzymes join each of the amino acids to the appropriate transfer RNA. (b) The messenger RNA and the transfer RNA for the first amino acid of the protein form a complex with the ribosome. Then theribsome effectively moves along themessenger RNA molecule. As it moves, the "anticodon" on the transfer RNA for each successive amino acid recognizes the corresponding codon on the messenger RNA and brings the amino acid into position for joining to the growing protein chain. In this way, the information encoded in the linear sequence of nucleotides in the DNA of the genes translates intothe linear sequence of amino acids in protein.***

Cells have an elaborate machinery for translating the nucleotide sequences in the DNA into amino acid sequences in proteins. It would be simple if amino acids could recognize their own codon and just line up in th right order on the surface of the DNA to be linked together by an enzyme, but nothing so simple evolved. Rather, cells use a different class of nucleic acids, RNAs (ribonucleic acids), to help translate the information contained in the DNA. RNAs are also strings of nucleotides, and consist of a sugar-phosphate backbones with a base attached to each sugar group, except that the sugar is ribose. Three of the bases in RNA are the same as in DNA (C, A, nad G) but instead of thymine (T), RNA uses uracil (U).

Gene transcription proceeds via the formation of a messenger RNA (mRNA) molecule. The mRNA carries the information that will specify the amino acid sequence from the nucleus, where the nucleotide sequence of the DNA contains the message, to the cytoplasm of the cell, where this message is translated and the protein synthesized. Translation of the message involves transfer RNA (tRNA) molecules as well as ribosomes, specialized structures on whose surface new proteins are assembled. As the amino acid chain lengthens, it begins to fold up, first in a helical configuration, and then in a three-dimensional globular shape.

During transcription, the two strands of the DNA separate and ribonucleotides assemble on one of the DNA strands in such a way that the bases will pair in the same way as for DNA replication, except that A (adenine) in the DNA is paired with U (uracil) in the RNA being synthesized. Thus, a sequence that is CGATC in DNA becomes GCUAG in RNA. When the mRMA reaches the cytoplasm, it binds to the surface of a ribosome where tRNAs with amino acids attached line up the amino acids in the correct order along the length of the mRNA, starting always with the amino acid methionine. A tRNA recognizes the codon in the mRNA by means of an anticodon, which aligns according to the base-pairing rules. As the tRNAs with their respective amino acids line up on the mRNA one by one, the amino acids link together by an enzymatic reaction and the tRNAs are released to react with new amino acids.

By studying the inheritance of all-or-none variation in peas, Mendel discovered how characteristics are transmitted from one generation to the next.

More than 130 years ago, Gregor Mendel, a Moldavian monk, formulated the two fundamental laws that govern heredity. People has suggested various ideas about heredity before Mendel began his experiments with peas, but none of these was supproted by experiments. When Mendel began his work, most breeders agreed that both sexes contributed to a new individual. They thought that parents of different appearance always produced offspring with an intermediate appearance. Thus, a cross between a plant with red flowers would lead to only plants with pink flowers. When future generations produced red and white flowers, breeders mistook this to mean that the genetic material was unstable.

Mendel conducted a series of experiments in which he crossed two varieties of pea plants with constrasting characteristics such as white and red flowers or round and wrinkled seeds. Remarkable, the characteristics he used can still be found today, precisely because they all have all-or-none variation. He published his observations in 1865 in a local scientific journal, but his findings went largely unnoticed until other scientists made the same observations 35 years later.

Mendel, who was very familiar with peas, knew that they always breed true (the offspring are exactly like the parent plant and like each other), because peas are self-fertilizers and the pollen fertilizes the pistil (female reproductive organ) even before the flowers opens. However, by cutting away the anthers (male reproductive organ) and brushing on pollen from another plant, Mendel figured out that peas could be cross-pollinated. Mendel had many types of true-breeding peas growing in his garden, and he wanted to know what would happen if he cross-fertilized lines of his peas with constrasting characteristics. One line of pea always produced smooth, round peas, but another one produced wrinkled peas. When the latter peas dried out at the end of seed maturation, some of the inner tissues collapsed, providing the seed with a wrinkled appearance (phenotype).

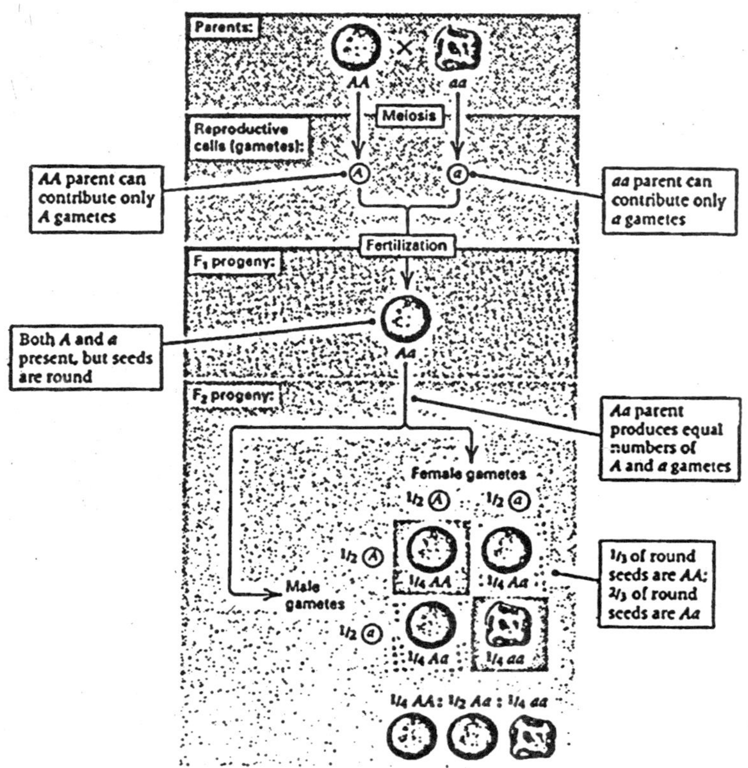

Figure 10.8: ***Inheritance of the round and wrinkled characters in peas. When plants that always produce round seeds are mated with plants that always produce wrinkled seeds, the plants in the first generation (F_1) all produce round peas. However, the F_1 plants still have the wrinkled gene. When the F_1 plants are crossed with one another, one quarter of the seeds in the F_2 generation are wrinkled and three quarters are found. This patern of inheritance can be explained by assuming that each plant carries two copies of the genes for each character, and that being round is dominant over being wrinkled. The original parents, which breed true, are either AA or aa, and when these plants are crossed, gametes are either A or a, and all off-spring are Aa. When these F_1 plants are crossed, both can produce A and a gametes and there are more possibilities for the offspring: there will be two Aa plants for every one that is either aa or AA. However, because A is dominant, three quarters will be round (AA and Aa) andonly one quarter will be wrinkled (aa).***

When Mendel crossed round and wrinkled peas, he observed that the first generation (the F_1 generation) consisted entirely of round peas. When these round peas were allowed to sprout, grow, and flower, and when the plants were allowed to set seed, most seeds of the F_2 generation (about 75%) were round and a minority (about 25%) were wrinkled. None were in between; none were just a little wrinkled or nearly round. Thus a characteristics that disappeared in the first generation reappeared in the second.

Mendel repeated these experiments using seven other discontinuous characterisitcs (for example, green and yellow seeds) and confirmed his experiments with round and wrinkled peas. Interestingly, he found that these seven different characteristics were not linked, but were transmitted to the next generation independently. Thus, when he crossed a round yellow pea with a wrinkled green pea, the first generation has only round yellow peas. However, in the second generation he found four types: the two original types and two additional combinations of the characteristics (round green and wrinkled yellow).

From such experiments, Mendel drew two important conclusions. First, characteristics or traits transmit to the next generation as discrete units, now called *genes*. Second, an individual must contain two copies of each of these units, and each parent transmits only one copy to the next generation. That is the only way Mendel could account for the disappearance and subsequent reappearance of a characteristic. The implication is that the unit (gene) is always present, but may not be expressed, as is the case with the wrinkled or green characteristics in the first generation of crosses of round and wrinkled or green and yellow peas.

Although Charles Darwin and Mendel were contemporaries (Mendel published his work in 1865, fifteen years before Darwin's death), Darwin did not know of Mendel's work. Darwin knew that characteristics could disappear and reappear, but he did not understand their mode of inheritance. In his book, *On the Origin of Species by Means of Natural Selection*, Darwin discussed the notion that a specific trait can be inherited by one child but not by another and that traits of grandparents sometimes appear in the grandchildren although they did not appear in the parents. He did not understand this skipping of a generation, which Mendel's experiments so beautifully explained.

Why are there smooth and wrinkled peas? Biochemist have recently discovered that wrinkled peas lack one of the important enzymes for starch synthesis. During their development, peas import sucrose from the rest of the plant. This sucrose is quickly converted to starch, which makes up 60% of the weight of a mature pea seed. Wrinkled peas lack one of the enzymes for starch synthesis, so when the seeds mature they contain a lot of sucrose and water and much less starch than normal. Then, when the seeds dry out, they wrinkle. The absence of the enzyme is caused by an alteration in the gene that encodes the information for synthesizing this enzymes. A single gene usually controls all-ornone variation, as in the case of the smooth and wrinkled peas.

Many agronomically important characteristics, such as yield or protein content of seed, show continuous variation, and the inheritance of such traits is controlled by many genes. Such traits are referred to as *multigenic* or *quantitative traits*. Thus, the ability of a plant to take up soil nutrients, photosynthesize, transport photosynthate to the seeds, and withstand drought all affect yield. And each of these characteristics is controlled by many genes, making yield truly a multigene trait.

Most organisms have two copies of every gene in every cell, except in the sex cells (sperm and egg cells), which each only have one copy of each gene. In humans, each cell has 22 pairs of chromosomes and 2 sex chromosomes, for a total of 46. Maize has 10 pairs of chromosomes and *Haplopappus*, a plant that thrives in dry areas, has only 2 pairs. Because cell have two copies of every chromosome, they also have two copies of every gene, one copy on each of the chromosomes that make up a pair of chromosomes. When these two copies are identical, the organism is said to be *homozygous* for that gene. If one of the two gene copies has mutated, they will be different, or *heterozygous*. These different forms of the gene are called *alleles*. Plants that normally self-fertilize, such as peas and beans, are homozygous for a very large number of genes, whereas plants that normally outcross, such as maize; may be heterozygous for most of their genes.

Mitosis and Meiosis are Important Cell Processes During Which Genetic Information is Passed on from Cell to Cell or Generation to Generation.

As noted earlier, genes are located on filamentous structures, called *chromosomes*, in the cell nucleus. Most complex organisms such as flowering plants, insects or mammals have in their cells some 20,000-60,000 genes but only a small number of chromosomes. Thus thousands of genes are lineraly arranged on each chromosome. The chromosomes play a vital role in passing on genes, and therefore phenotype characterisics, from a mother cell to the daughter cells during cell division, and from the parents to the offspring during reproduction. Cell division, which in plants occurs in meristems, is preceded by mitosis, a process of chromosome duplication, and the subsequent separation of the two sets of chromosomes. Thus for a short while a cell contains four copies of every gene, but as the new chromosomes separate, each daughter cell again contains two copies of every gene. During the formation of egg cells and sperm cells, a different type of cell division occurs. A single cell undergoes one round of chromosome replication followed by two round of chromosome separation and cell division. Four cells are formed in this process, which is called *meiosis*, and each cell ends up with a single copy of each gene. If the organism is heterozygous for that gene, then half the sex cells will have one allele of the gene and the other half will have the other allele.

When two sex cells fuse during fertilization to start a new organism, the cells of this new organism will again contain two copies of every gene. A plant homozygous for a given gene will produce sex cell, all of which have identical forms of that gene. That is in terms of that gene, it produces only one type of reproductive cell. However, a plant that is heterozygous for a certain gene, with (for example) one normal and one mutant copy, will produce two types of reproductive cells. Half the sex cells will carry the normal gene, half will carry the mutant gene. The fact that many plants are normally heterozygous for many genes creates many possibilities for variation in the offspring, because random assortments of genes are brought together when the sperm cell fertilizes the egg cell.

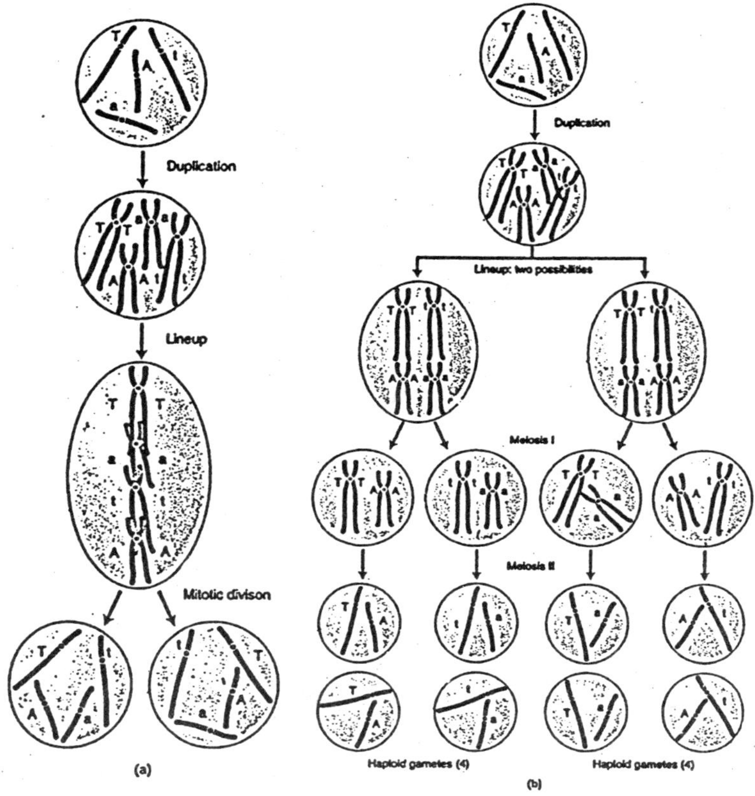

Figure 10.9: ***Mitotic and meiotic cell divisions. (a) The mitotic methods of cell reproduction occurs in plant meristems. One member of each gene pair (e.g., T and t) located on one member of a homologous chromosome pair (e.g., the large chromosome). One of the pair came from the male gamete (e.g., the blud chromosome) and the other from the female gamete (e.g., the red chromosome). Note that in mitosis two cells are formed, each one genetically identical to the original cell. (b) The meiotic method of cell reproduction results in the formation of sex cells (pollen or egg cells). Note that four haploid gametes are formed from a diploid cell. The key is the separationo of the duplicated homologous chromosome pairs during meiosis I and the separation of each member of the duplicated chromosomes during meiosis II. Depending on how the homologous pairs line up during meiosis I, different combinations of genes***

end up in the haploid gametes that are formed.

Let us consider a plant with a haploid chromosome number of 2. Its normal diploid cells will have two pairs of chromosomes in each cell. T and t are alleles of the same gene and are located on the long chromosome (note that T came from the male parent and t from the female). A and a are alleles of a different gene and are located on the short chromosome. Before mitosis begins, the cell has four chromosomes in its nucleus: a pair of each of two types. Mitosis involves three steps:

- Duplication of the chromosome genetic material (DNA); this acutally occurs during the time that the cell nucleus is visible and chromosome cannot be seen as distinct structure.
- Lining up of the duplicated chromosomes, usually at the center of the cell.
- Separation of each member of the duplicated chromosome are genetically identical, such that each of the two new cells gets one copy of the pair.

Because the two members of the duplicated chromosome are genetically identical, the two new cells that are formed contain the same genes (T and t, A and a) as the cell that produced them.

The earliest part of meiosis is identical to mitosis: The chromosomes are *duplicated* and each one becomes a double strand. The chromosomes *line up* in the center of the cell; however, in this case, they line as homologous pairs. Now, *separation* occurs such that the duplicated chromosomes separate from each other, one duplicated member going into each of two new cells. The resulting cells clearly have only half of the number of chromosomes (in this case, 2) as the originating cell (4). These two chromosomes are double, so that in meiosis II, the members of the doublet separate, just like in mitosis. Now there are four cells formed, each with a haploid chromosome number (2).

An additional feature of meiosis that is essential for genetic variability. At the lineup and separation stages of meiosis l, the only requirement is that each new cell gets one of the two members of the homologous pair. So it is equally probable that one of the new cells will get the "A" and "T" chromosomes and the other "a" and "t" as it is for them to get the "A" along with "t" and "a" with "T". So in our simple case, there are four possible chromosome combinations in the gametes: AT, at, At, and aT.

Few plants have just two pairs of chromosomes (diploid number of 2). For example, corn has 10. As we have just seen, the number of possible gametes formed with two chromosome pairs, just with regard to paternally and maternally derived chromosomes, is 4 (2^2). For 10 chromosomes, the number is 1,024 (2^{10}). If we consider that there are often allelic differences between the two chromosomes (see our example, T and t, and A and a), there are 1,024 genetically different gametes possible. These combinations mean that in sexual reproduction, the offspring will not be genetically identical to the parents. This provides much genetic variation, the raw material of evolution by natural selection. It also provides a challenge to plant breeders, who seek to minimize variation in offspring. Additional genetic variation is created when chromosomes exchange portions during crossing over in meiosis.

MULTIPARTITE NATURE OF GENES

Orderly development of an organism requires that the correct proteins be synthesized at the correct time and place during development, and in response to the proper stimuli. For example, the enzymes that causes softening of cell walls in the ripening fruit are only synthesized in the fruit and as a result of an increase of the hormone ethylene.

So far, we have described genes only in terms of the information they contain to specify a sequence of amino acids and the structure of a protein. This part of a gene is called the *protein-coding* region. However, a gene is more complex (much longer, really) than just a protein-coding string of nucleotides. First of all, the protein-coding portions, called *exons*, may be interrupted from one to a dozen times by very short to very long stretches of DNA, called *introns*. Introns are transcribed when RNA is first made in the cell nucleus, but they are later removed when the initial product of transcription is processed into a mature mRNA that is transported to the cytoplasm. These interruptions in the coding sequence are characteristic and different for every gene, and can

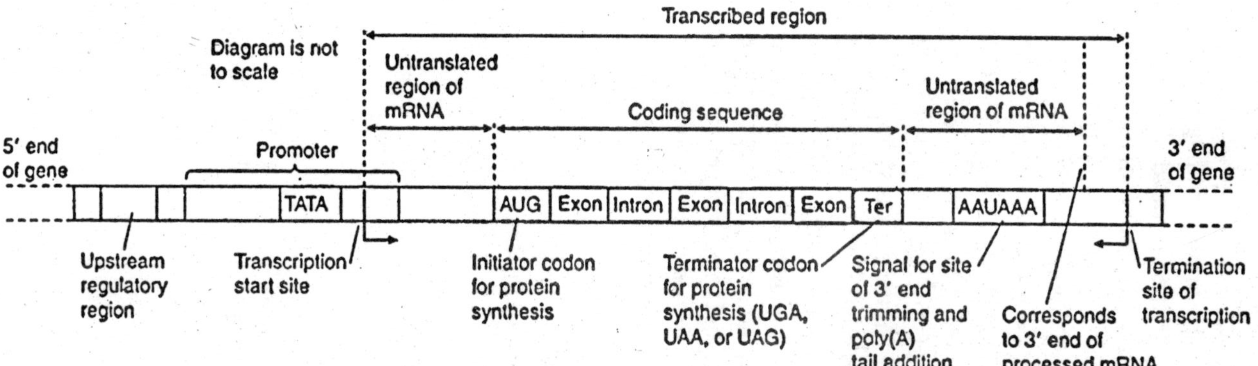

Figure 10.10: *Structure and organization of a eukaryotic gene. A gene has several sections. The transcribed region functions as a template for synthesizing RNA, which then translates into the protein product of the gene. The transcripbed region is interspersed with noncoding sequences that partition the region intocoding sections (exons) and noncoding sections (introns). The transcribed region is flanked n either side by noncoding sequences that play a role in regulating the gene. Most of the regulatory sequence elements motifs important for the "promotion" of transcription. The most highly conserved part of the promoter is the TATA box. which is usually located within the first 50 bp of thetranscription. The most highly conserved part of the promoter is the TATA box, which is usually located within the first 50 bp of the transcription start site. The TATA box coordinates the recruitment of RNA polymerase to the gene.*

also be involved in regulating the gene's expression. Second, a DNA segment on each side of the protein-coding region specifies when and where the gene is to be activated during development, to which stimuli (hormonal or environmental) the gene should respond, in which cells it should be active, how much mRNA is to be made, and when the gene is to be turned "on" and "off". These *regulatory regions* are, of course, of great importance for the orderly development of a cell and an organism.

Thus, genes that encode enzymes for the synthesis of chlorophyll are "on" in the leaves, but "off" in the roots and the flowers. Furthermore, they are "off" in the dark and "on" in the light. When a root finds itself in soil that is rich in nitrate, genes are turned "on" that encoded proteins needed to take up nitrate, its transformation into ammonia, and the use of this ammonia in amino acid biosynthesis. When the spore of a plant pathogen germinates on the surface of a leaf and tries to penetrate into the leaf cells, defense genes are rapidly turned on in the cells of resistant plant varieties. These defense genes encode enzymes that synthesize toxic compounds that will kill the invader. When light strikes a seedling that has been growing in the dark underneath the soil surface, hundred of genes that were completely off or barely on are turned on and the cells start making hundreds of new proteins that let the chloroplasts develop, and photosynthesis and autotrophic growth begin. These are only a few examples of the gene activity regulation in response to specific stimuli.

Regulating gene activity is the responsibility of proteins called *transcription factors* that bind to the regulatory regions of genes, and these transcription factors are themselves the products of genes. Thus, turning genes "on" and "off" is not a simple matter and may involve a regulatory cascade in which the product of gene A activates gene B, whose products activates gene C, whose product activates gene D by binding to its regulatory region.

Plant breeding and genetic engineering both involve transfer of genes from one organism to another. In plant breeding, the breeder selects for an ultimate outcome or a phenotype, thus ensuring that the whole gene regulation pathway will operate correctly. Let us take the example of transferring resistance to a specific pathogen from a wild relative of wheat to a cultivated wheat variety. A resistance gene may encode a protein that lets the plant detect the invader quickly so the plant can turn on its defenses. The gene that encodes this detector protein responds to a stimulus from the pathogen— perhaps a chemical or metabolite made by the pathogen— and this response involves several other genes. Thus when the breeder transfers resistance to that specific pathogen from a wild variety of wheat to a domesticated wheat by selecting for pathogen resistance in the field, the entire regulatory cascade must work correctly. If the new combination carries the resistance gene but is not correctly expressed, then the new variety would not resist the pathogen.

Genetic engineers transfer one gene at a time and need to understand how the gene they transfer is regulated. First, they need to transfer not only the protein-coding part of a gene but also its regulatory region. Or they can equip the gene that a new regulatory region that ensures correct expression. For example, regulatory regions of bacterial genes generally do not work in plants because the gene transcription machinery of bacteria differ from that of plants. Similarly, a regulatory region of the gene from a monocotyledon (maize) may not work in a dicotyledon (bean). Thus, a full understanding of the regulation of gene expression is of great interest not only to plant biologist who want to know how plant development is regulated, but also to genetic engineers who transfer genes between organisms.

MANIPULATION OF GENES IN THE LABORATORY

Since the 1960s, molecular geneticists have made tremendous advances that have led to the development of exciting new areas of science in DNA technology and plant biotechnology. Biotechnology is the use and manipulation of living organsims, or substances obtained from these organisms, to make products of value to humanity. Although *biotechnology* is a relatively new term, this idea is not new—people have bred plants and animals to express particular traits since the beginning of civilization. However, biotechnology based on manipulating DNA outside of the living cell is a newer and much more powerful application of the technology. These new techniques let scientists isolate genes from one organism and then insert them into another, to produce an organism with a new trait. Before these developments, plant breeders were largely limited to genetic exchanges within a new species and between closely related species. Now scientists can transfer genes (genetic engineering) and therefore inherited traits between very different organisms. This ability to transfer genes among humans, plants and bacteria has revolutionized biotechnology.

Progress in genetic engineering depends on a new technology that lets scientists isolate, identify, and clone (produce multiple copies of) genes. This recombinant DNA technology, called *gene cloning*, was developed in the 1980s, but originated with the discovery of plasmids and restriction enzymes in bacteria in the 1970s. Scientists make recombinant DNA (rDNA) by joining, or "recombining", DNA segments from different sources: plant and bacterial DNA, or plant and animal DNA. Because DNA strands from all organisms have the same chemical structure, they can be cut into segments and the segments linked together again in new and different ways.

Cutting the DNA is done with special enzymes called *restiction enzymes*, which occur in bacteria as part of a natural defense system against invading viruses. Restriction enzymes are highly specific to the nucleotide sequence, and each enzyme only cuts at one specific short base sequence in the DNA. For example, the enzyme *Eco*R1, found in the intestinal bacterium *Escherichia coli*, cuts only at GAATTC (the complementary strand, CTTAAG in this case, is assumed), and such a sequence occurs on the average only once every 4,000 nucleotides of DNA. An interesting feature of most restriction enzymes is that they make a staggered cut across the two strands of DNA, producing what molecular biologist call "sticky ends". Another enzyme, DNA ligase, can rejoin (or ligate) those sticky ends together again, and this can happen whether the DNA is from the same organism or from different organisms. Such manipulation of DNA is the laboratory are often called *gene splicing*.

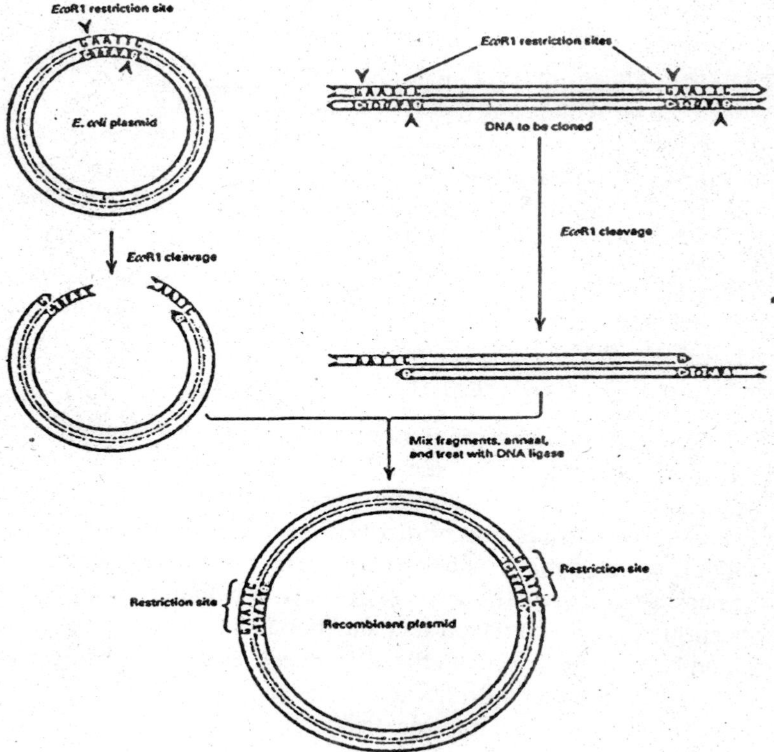

Figure 10.11: ***Construction of recombinant DNA molecules. A specific restriction enzyme—EcoR1, for example, as shown here—makes a staggered cut producing "sticky" ends. The single-stranded end of one fragment can therefore recognize and bind to the end of any other fragment produced by the same enzyme, even if the fragments originally came from the DNAs of different species. The two fragments can then be joined and circularized by the action of an enzyme called DNA ligase. Restricting and ligating DNA in this way forms the basis of recombinant DNA technology. The red arrowheads indicate the EcoR1 cleavage sites.***

Scientists carry out the job of cloning—of producing multiple copies of a gene or DNA segment—with the help of what is called a *vector*. The most common vectors are plasmids, and these are the delivery vehicles scientists use to introduce recombinant DNA into a host cell. Plasmids are (small) double-stranded circular DNA molecules that replicate in bacteria. An important property of plasmids is that they can sometimes be transferred from one organism to another. By using a restriction enzyme, a scientist can cut the circular plasmid open. If DNA from another organism has been cut with the same restriction enzyme so that the same sticky ends were created, DNA ligase can be used to reform circular molecules that contain a segment of foreign DNA spliced into the site where the plasmid was cut. The new plasmid is transferred to bacteria that have been specially treated for this purpose; when the bacteria are allowed to multiply, the plasmids are copied within the bacteria. Million of copies of plasmid and the inserted gene can be reisolated from these bacterial cells.

PLANT REGENERATION TECHNOLOGY

When small pieces of plant tissue are put in sterile culture on a solid nutrient medium, the cells proliferate and a callus forms, and is auxin and cytokinin are present in the correct amounts, shoots will form. Shoots normally arise from small groups of rapidly dividing cells within the callus. The discovery led scientists to ask a fundamental question: Can a whole plant be grown from a single cell, given the correct nutritional and hormonal environment? Many years ago, German, Japanese, and U.S. scientists discovered independently that this is indeed the case, and this discovery let them draw important conclusions about genes and how they function.

Making an entire plant depends on the correct expression of at least 10,000 genes, perhaps more. When a living plant cell is isolated from a mature tissue, it can be induced to start dividing again and all the genes necessary to make an entire organism can be induced to function again in the correct sequence. This ability of a singly mature plant cell to give rise to an entire organism is called *totipotency*. Because plant cells are interconnected by their cell walls, in a kind of honeycomb, it is not so easy to isolate single cells. It is much easier to digest the cell walls of a small piece of plant tissue with enzymes and then isolate the protoplasts (naked cells without walls). The protoplasts are quite fragile, but they too can be cultured and regenerated into entire plants.

Regenerating whole plants from single cells (whether a protoplast or a cell that is still part of a tissue) is an important aspect of genetic engineering. Genetic engineering involves introducing a gene, usually from a different plant or an unrelated organism such as a bacterium, into a plant. If you can get the gene incorporated into the genome of just one cell and then encourage the cell to divide and form a whole organism, then you can get that gene into all the cells of that new plant. Two methods of introducing a new gene are commonly used, *Agrobacterium*-mediated gene transfer or a gene gun. With either method only a few cells are transformed; most cells remain untransformed. The later must be prevented from growing into whole plants, otherwise you won't know which new plants are transformed and which are not.

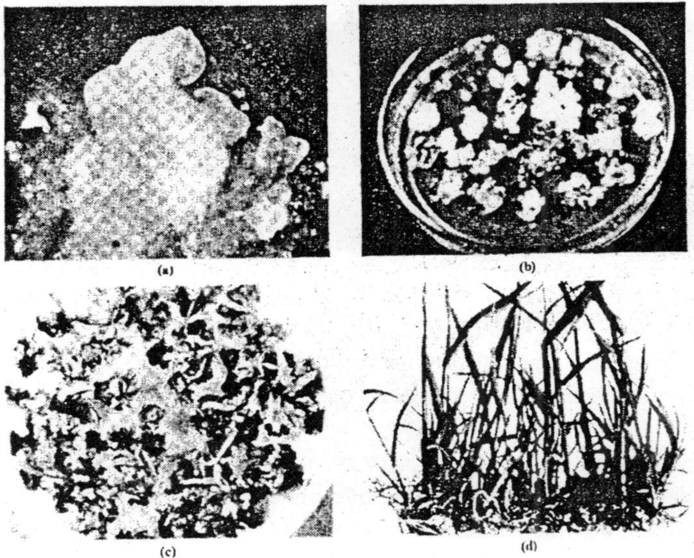

Figure 10.12: ***Embryogenic callus and plant regeneration. For most plants, the levels of two hormones, auxin and cytokinin, in the medium and the ratioof one to the other determine whether a small piece of tissue only forms callus (when both hormons are at high levels), sends out shoots (high cytokinin, low auxin), or sends out roots (high auxin, low cytkinin). (a) Proliferation of embryogenic callus derived from young leaf tissue. (b) Shots began to regeneratte from pieces of callus after the biologist reduced the concentration of cytokinin in the media. (c) When shots have elongated, the biologist cut them off the callus and placed them on a media with a high auxin concentration. (d) Roots develop on media with a high auxin concentration. In this example, the plant is sugarcane.***

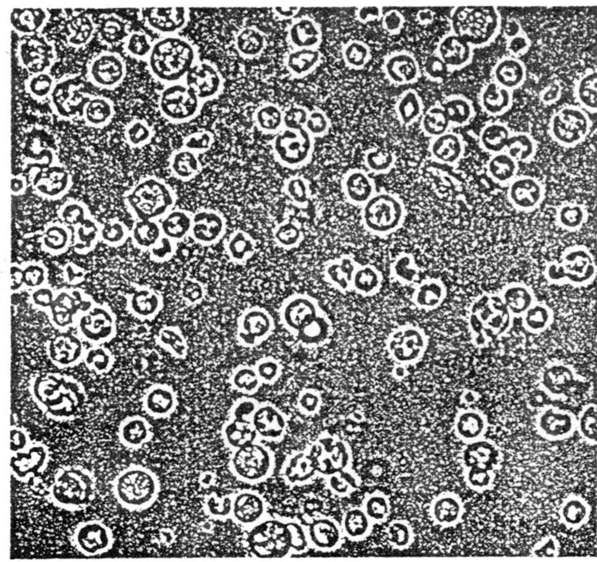

Figure 10.13: ***Freshly isolated protoplasts. Plant cells from whichthe cell wall has been removed are called protoplasts. The most common procedure to make protoplasts involves enzymatic cell wall degradation using a mixture of enzymes extracted from fungi, snail gut, an termite gut (all organisms that can digest cell walls). The protoplasts shown here were isolated from citrus leaves.***

To kill all the untransformed cells, scientists use a herbicide or an antibiotic in the culture medium, and let the transformed cells survive by linking them to a second gene that inactivates the herbicide or antibiotic as soon as it enters the cell. This gene is called the *selectable marker*. Genetic engineers usually introduce two genes at the same time: the gene that encodes the novel trait they wish to introduce—also called the *gene of interest* and the selectable marker gene. Molecular manipulations later eliminate the molecular marker gene. However, during the tissue culture phase the selectable marker gene lets the cells synthesize the enzyme that breaks down the antibiotic or herbicide, so only transformed cells survive and generate transformed plants. Plants produced with these techniques are usually called *genetically modified* (GE) or *genetically enhanced* (*GE) plants*. We use this terminology here for consistency recognizing that plants improved by classical breeding are also genetically modified or enhanced.

INTRODUCTION OF DNA USING A "GENE GUN"

The goal of GM technology is to isolate one or more specific genes and introduce these into plants. GM crops (or foods) are crops produced by using molecular techniques to introduce new genes, followed by several years of plant breeding. For many plant species, one or a few genes can be introduced via the natural gene transfer system of the pathogenic coil bacterium *Agrobacterium tumefaciens*, which causes tumors (called *crown galls*) in many plants. When these bacteria infect a wound site, usually on the stem close to the ground, the infection disturbs the normal healing process. Instead of making a protective tissue that covers the wound, cells proliferate into cancerous growth. Cells from this tumor can be grown in tissue culture, and unlike normal plant cells, they continuously proliferate even when hormones are absent from the culture medium.

Molecular biologists in Belgium, the Netherlands, and the United States found that after the bacteria attach themselves to the wound site, they transfer some of the genes on the Ti plasmid to the plant cells. The bacteria have a large plasmid of 200,000 bases (200 kilobases or 200kb) (also called the *Ti plasmid*, for "tumor inducing"), and when the bacteria infect a wound site a portion of the plasmid, called the *T-DNA*, it cut out be enzymes, coated with proteins and then becomes integrated in a plant chromosome. This integration is random, meaning that scientists cannot yet direct the T-DNA to a specific plant chromosomal location.

Genes thus transferred carry the in formation necessary to make auxin and cytokinin. The abnormally high levels of these hormones produced in the plant cells prompt cancerous cell growth. This resembles the situation in culture where the continuous availability of the same two hormones in the culture medium causes cell proliferation and callus formation. After the genes have been transferred from a single bacterium to a single plant cell and integrated into a chromosome of the plant cell, the hormones in the culture medium induce the cell—which is now transformed with the bacterial genes—to proliferate. So the bacterial genes will be

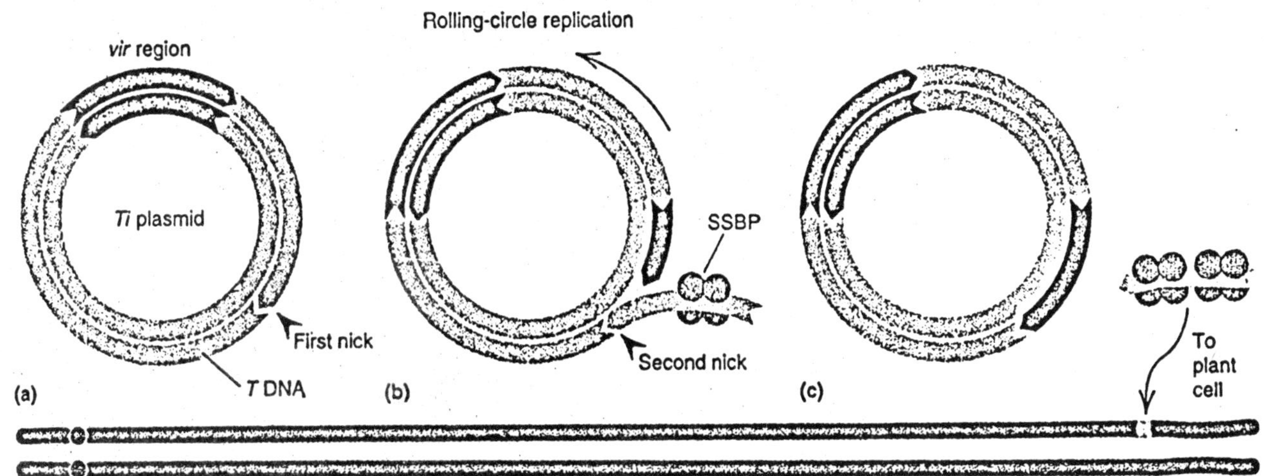

Figure 10.14: *Genetic transformation of a plant genome by T-DNA from the Ti plasmid. (a) A nick forms at one end of the T-DNA aftr the bacteria have attached to a wound ite on the plant. (b) Replication elongates one end and displaces the other end where the first nick was formed. The single stranded binding protein (SSBP) that is coded by one of the vir gens then stabilizes this displaced piece of single-strandd DNA. A second nick terminates replication. (c) The single-stranded DNA that is coated with the SSBP is transferred into a plant cell and integrates into the plant genome.*

present in every cell derived from the transformed cell. This transfer of DNA between the different species is sometimes called "promiscuous", because DNA is usually transferred only between individuals of the same species, as part of the process of sexual reproduction. Nevertheless, this transfer of DNA between totally unrelated organisms is a natural process. Soon molecular biologist realized that if they substituted other genes for the genes that the bacterium transferred to the plant cell, they could obtain transformed plant cells that carried any gene they wished to introduce into the plant.

Not all types of plants can be easily transformed with *Agrobacterium* (especially the monocots), to researchers have developed other methods to introduce foreign DNA into the genomes of important crops such as maize, wheat, sorghum, sugarcane, beans, and soybean. John C. Sanford and his colleagues at Cornell University developed a microprojectile gun (or "gene gun") to shoot DNA into plant cells directly, through the cell wall and the cell membrane particles with DNA and constructed a gun that accelerated the particels with enough speed to enter the first cell layer of a tissue. For reasons not yet completely clear, DNA that enters the cells in this way is integrated into the chromosome and passed on the cells' progeny. By using tissue culture techniques, as describe earlier, researchers can generate a whole new plant from this transformed cell.

GENOMICS IMPROVEMENT OF CROP

Gene technologies and tissue culture techniques have dramatically increased the possibilities for crop improvement. It is no longer necessary to rely only on the genes of close relatives of a particular crop because biologists can use genes from any source and can introduce a particularly useful gene into many crops. Using the same bacterial gene, for example, they can make 10 different crops tolerate the same herbicide. In the 1990s a new science emerged: genomics, or the study of all the genes of an organism. Genomics is concerned with the nucleotide sequence of all the genes, the functions of all the genes, and the expression (when, where, in response to which stimuli) of all the genes. Scientists use high-throughput methods to analyze 20,000 or more genes at once!

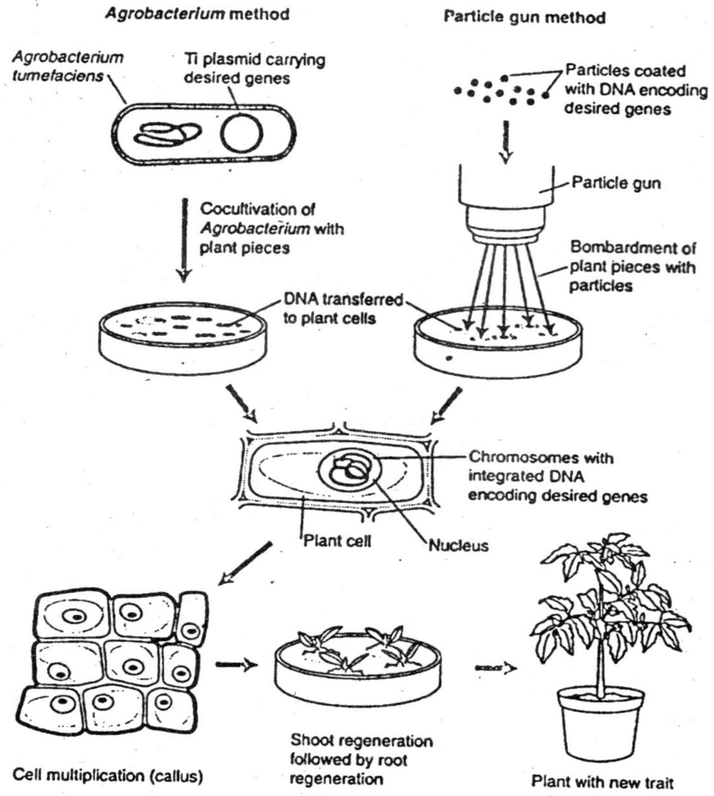

Figure 10.15: ***Schematic representation of two different ways tocreate transgenic plants.***

In the Agrobacterium method (left), the bilogist inserts DNA carrying the desired genes into the tumorinducing (Ti) plasmid of the bacterium, and when the bacterium infects a wounded tissue, it transfers this DNA toa cell nucleus and intergrates it into the chromosome. In the particle gun method, metal particles coated with DNA become integrated into the plant chromosome. When a new plant regenerates from a single transformed cell, all the cells in the plant carry the new genes.

Advances in DNA-sequencing technology and computer science have now made it possible to produce tens of thousands of random insertion mutants that allow scientists to study the characteristics of plants in which genes are inactivated by the insertion of a large piece of DNA. Such a collection of "knockout" mutants with a different gene inactivated in every plant constitutes a valuable resource for molecular geneticists. A collection of such mutants is already available for *Arabidopsis* and may be available for rice as soon as 2005. Agricultural biotechnology companies are already racing to produce such collections or rice plants.

Another new technology, called *expression profiling*, allows scientist to determine the level of gene expression (amount of mRNA) of all the genes at once and to study the effect of environmental perturbations (lack of nutrients, water deficit, shade, pathogen attack) on gene expression. This ability will quickly lead molecular biologists to manipulate the pathways of gene expression, either with new biodegradable chemicals or by subtly tering a few genes of the plant. For example, a plant may resist a pathogenic fungus because it can quickly mount a defense response by inducing expression of several defense genes. A normal plant susceptible to fungus can't defend itself fast enough. Once scientists understan which gene is the master switch for this response, they may be able to speed up the response, allowing the plant to mount an effective defense. This type of genetic engineering will require subtle changes in genes and will resemble the normal process of evolution more closely than do current genetic technologies that introduce completely new genes.

11

Genes in Growth and Development

The molecular basis of crop improvement and noted that specific, traits are encoded in genes, short segments of DNA that can be transformed from one organism to another. The expression of genes is highly regulated and changes dramatically during the plant's development and as a result of environmental signals. To understand how scientists might change plants to suit human needs as food sources, you first need to understand plants. That fertilized egg cell that constitutes the beginning of the life of every plant already contains all the 20,000 to 50,000 genes the plant will ever have! Their orderly expression controls development, which involves formation of organ systems such as roots and shoots. Such development requires integrated cell division, which gives rise to more cells, and cell differentiation, which gives rise to different cell types as new organs, each with multiple tissues, form. During this process many genes respond to environmental stimuli, changing the course of development. In addition, the plants establish defense mechanisms against pathogens, insects, and other pests. These mechanisms consist of proteins or other molecules, the information for which is encoded in the genome. Understanding how genes regulate plant development will help you gain insights into what scientist might accomplish by manipulating the expression of specifc genes to transform crops into more efficient food products.

Each type of cell in a tissue has its own specific set of protein, required to carry out the unique functions of that cell, and each protein is encoded by a gene. Many so-called housekeeping genese are expressed in all cells and at all times, but thousands of genes are activated or up-regulated and down-regulated at specific stages of plant development. How the plant regultes the orderly expression of these thousands of genes greatly interests developmental biologists. When a plant develops from an embryo, a seed forms first; after seed germination, the vegetative body (roots and shoot) forms; finally, the appearance of flowers initiates the reproductive phase. Each stage requires the activation of new genes.

The expression of genes during development is influenced by such environmental signals as light, the availability of minerals, or a period of low temperature, and is coordinated by internal signaling molecules. These signals can be small molecules such as hormones that circulate within the plant, or proteins that pass from one cell to the next. Such signaling molecules help cells or organs communicate with one another and with meristems, the sites where cell division occurs and new organs are initiated. Cell differentiation is accompanied by the elaboration of the biosynthetic machineries that enable the plant to carry out photosynthesis, take up nutrients from the soil, transport sugars, and store protein and starch in special structures. The integrated, orderly expression of all this genetic information stored in the cell nuclues (and to a small extent in the chloroplasts) is truly awe-inspiring.

CELLS, TISSUES, AND ORGANS

The vegetative body of all plants, in all their apparent diversity, consists of two parts: a root system and a shoot system. The root system consists of a primary root with secondary and tertiary roots: the shoot system has a primary stem with branches, all of which usually carry leaves and may produce flowers. The arrangement of these major organs (roots, stems, leaves) in young plants and belonging to the two major groups, the monocots and the dicots. The main function of the shoot system are photosythesis and the formation of the reproductive organs. A stem consists of a number of segments,. called *internodes*, connected by nodes. In a young plant a leaf is attached at each node. The functions of the root system are to take up water and minerals, to store starch, and to anchor the plant in soil.

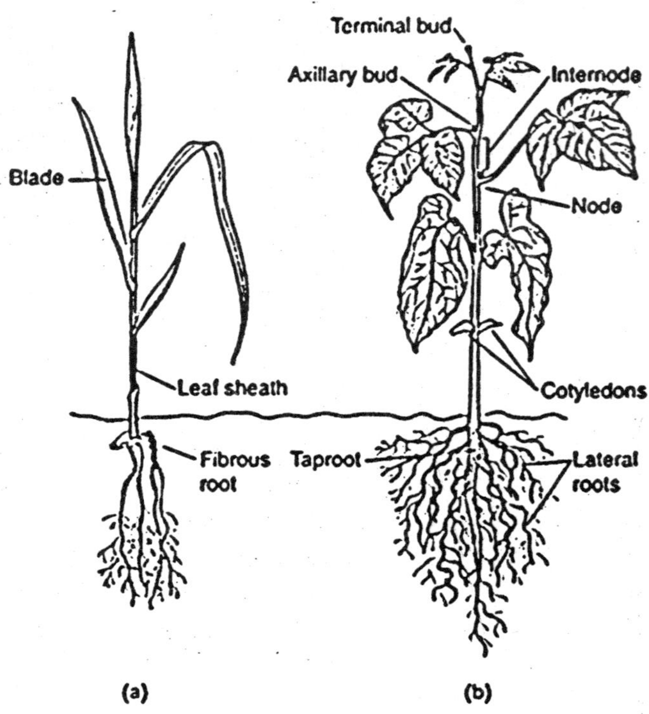

Figure 11.1: ***The organs of young vegetative monocot (wheat) and dicot (bean) plants. (a) Monocots have elongated leaves with parallel veins (vascular bundles) and a system of numerous parallel roots; grasses, cereals, and palms are monocots. (b) Dicots have broad leaves with branded veins and a branched taproot with lateral roots. Most common plants, such as soybeans and other legumes, potatoes, tomatoes, and cotton,are dicots.***

Plant have four reproductive organs, all contained in the flower: sepals, petals, stamens (male reproductive organs), and carpels (female reproductive organs). As in animals, each plant organ consist of several tissues and each tissue contains several cell types.

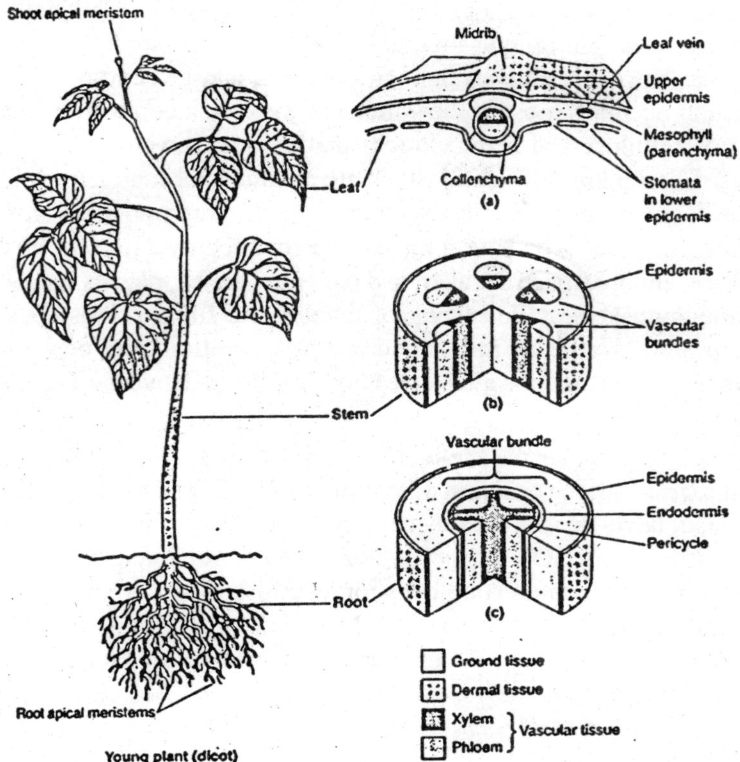

Figure 11.2: ***The body of a plant showing the organization of the three tissue systems in the organs. The cross sections through each organ shown the different organizations of the three tissue systems in the leaf (a) stem (b) and root (c) Each organ has vascular tissues (xylem and phloem), dermal tissue (epidermis), and ground tissue (mesophyll, cortex, and pith).***

Plant organs gave three types of tissues—dermal tissues on their surface, conductive tissues, and ground tissues—and their locations in a dicot plant are shown in. All together these tissues contain some 40 different cell types that are distinguishable on the basis of their morphology and function. The human body, in contrast, contains hundreds of different cell types indicating that plants are simpler organisms.

Cells are the basic building blocks of tissues, and the cells of all organisms contain more or less the same internal (subcellular) structures (organelles). Compared to animal cells, plant cells have at least four unique structures: large chloroplasts for photosynthesis or other plastids in nongreen cells, a large central vacuole that occupies most of the volume of the cell, a cell wall made out of cellulose and hemicellulose that encases the entire protoplasm, and plasmodesmata, which are channels through the wall that connect one protoplast with another. In common with animal cells, plant cells have a nucleus surrounded by a nuclear envelope, endoplasmic reticulum and Golgi apparatus, plasma membrane, cytoskeleton consisting of tubules and fibrils, ribosomes for protein synthesis, mitochondria for respiration, and peroxisomes where reactions are carried out that produce hydrogen peroxide.

Each subcellular organelle has its own unique set of proteins that carries out the function of that organelle. For example, chloroplasts have the enzymes necessary for photosynthesizing and for assimilating CO_2. The same enzymes are not found in other organelles. One interesting question in cell biology is how proteins find their correct destination. Nearly all proteins are synthesized in the cytoplasm, yet they end up in about 25 different locations in the cell! Each protein contains its own address label, and the cell has machinery ("mail slots") that inserts the correct protein in the correct organelle.

Certain cells that are specialized for storage, as in seeds, roots, and tubers, also have unique organelles specifically to sequestor protein (protein storage vacuoles or protein bodies), starch (amyloplasts), or oil (oil bodies). Such storage cells do not have a large central vacuole, and most of the cellular volume is taken up by these storage organelles.

PLANT DEVELOPMENT BY REPETITIVE ORGAN FORMATION

There is a fundamental difference in the way plants and animals develop. Plants have specialized tissues, called *meristems*, that can continuously give rise to new cells and new organs. This gives plants a rather simple way of solving the problems of aging: when an organ is getting old, the plants allows it to die and makes a new one. Leaf shedding is a typical example: When leaves are young, they photosynthesize vigorously, but as they get older their metabolic rate declines. Eventually they age and die. Perennial deciduous plants make new leaves in each growing season. When annual plants are crowded, as in a field of wheat, new leaves appear at the top of the plant while older leaves near the bottom are dying. This ability of repetitively make new organs also permits plants to recover from major damage to the plant body (as done by grazing animals or insects).

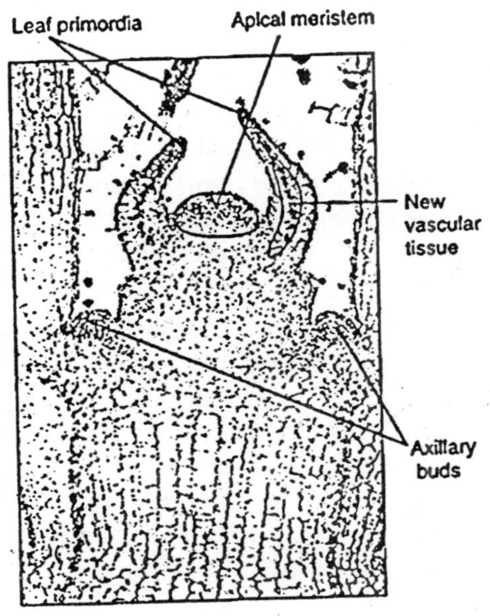

Figure 11.3: ***Longitudinal section through a shoot opex showing the location of the opical meristerm, two leaf primordia, and bud primordia (partially formed axillary buds).***

At the tip of each shoot is a shoot apical meristem (SAM) where the cells continuously divide. This is a very small, dome-shaped mass of cells. The SAM cells are the progenitors of all cells in the shoot. The tissue just below the SAM is also meristematic: The cells are dividing and contribute of the formation of new organs, especially the stem. Small protrusions of dividing cells, called *primordia*, arise on the surface of the SAM and just below it, and one by one these primordia grow out into small leaves. At the same time new meristems form, becoming axillary buds that remain quiescent until hormones trigger them into activity. Together, the apical meristem, the primordia, and the tiny leaves make up the apical bud you can find at the end of every twig. Apical buds also appear at the ends of other shoots, but are not always as easily identified as in twigs. The environment (temperature and day length), which causes changes in the cells' hormone level, regulates the activity of the SAM. These hormones in turn may activate gene cascades that ultimately control all the events in a meristem: rate of cell division, meristem size, exact position where a leaf primordium forms, and rate at which that primordium grows. (In addition to these apical meristems, cereals and other grasses also have growth zones called *intercalary meristems* just above each node in the stems. These growth zones explain why grass shoots up rapidly after being cut). Manipulating these genes may allow scientists to change the course of plant development.

The root apical meristem (RAM) is present near the tip of every root, just behind the roop cap, a small structure that protects the apical meristem as the root grows through the soil. The root cap has its own stem cells or initials, which continue to produce root cap cells. Root cap cells shed as the root grows, so the root cap never grows larger. The root itself originates from just a few (three or six) initials or stem cells. As a root elongates, lateral roots form not from the RAM at the tip of the root, but from meristems that are newly initiated somewhat back from the tip and below the surface of the root, in a tissue called the *pericycle* (Figure 11.4a). The pericyle can initiate new meristems when the hormone auxin triggers it to do so. As in stems, the pattern of root development is one of continued meristem and organ formation.

The SAM and RAM cells are small, about 100 to 1,000 times smaller than parenchyma cells, which make up the bulk of a young plant. Meristematic cells have a prominent nucleus, a dense cytoplasm full of ribosomes, and relatively undeveloped plastids. The main function of these cells is to divide and give rise to more cells, and this they do with amazing regularity—once every 36 to 48 hours. Meristematic cells have a thin, flexible cell wall, and after nuclear division a new cell wall forms separating the two new nuclei. At the periphery of the meristem, the cells begin to escape the siganls that keep them continuously dividing. Now the cells begin to enlarge enormously, mostly because of the increased vacuolar volume. Enlargement is accompanied by differentiation and by the expression of thousands of new genes that are not, or only barely, expressed in meristematic cells.

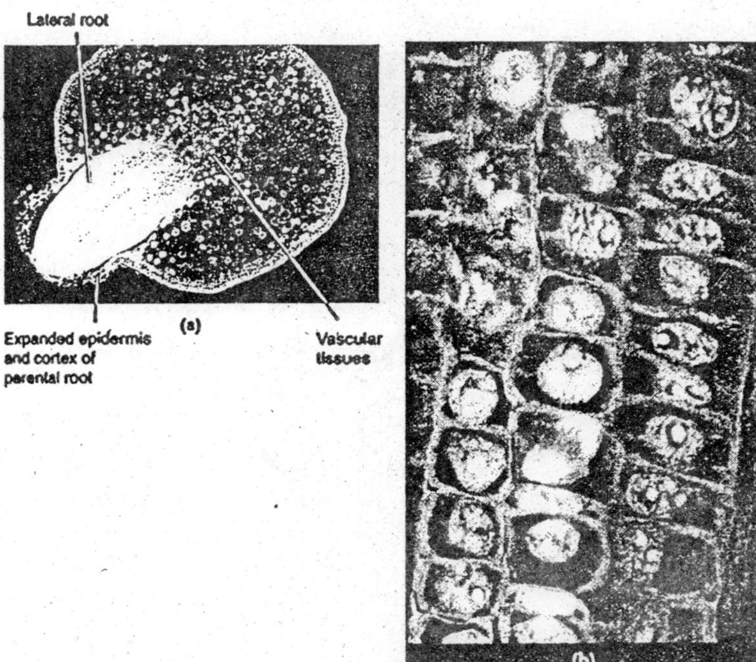

Figure 11.4: ***(a) Initiation of a lateral root in the pericycle of the main root. The hormone auxin triggers new cell divisions in the pericycle, and a meristem is formed. When the cells begin to enlarge, they puch the meristem forward, breaking through the cortex and the epidemics. (L) Portion of the root meristem of a hyacinth. Here, two cells are in mitosis, the process in which the chromosomes duplicate and are partitioned to the two daughter cells.***

The function of the SAM is to produce the cells that will make up the shoot system of leaves and stems. However, given the right environmental and/or hormonal stimuli, this apical meristem can be converted to a meristem that makes floral primordia. These floral primordia then grow out into flower organs (sepals, petals, stamens, and carpels). At that point, meristem activity ceases, and whatever cells remain stop dividing. The conversion of vegetative meristem to floral meristem is also controlled by a series of genes, as is the differentiation of the floral organs. In many plants, leaves, sepals, and petals have a very similar morphology yet are unique in function and in the proteins they contain. This uniqueness is caused by the expression of specific genes.

ENVIRONMENTAL STIMULI AND HORMONES PLANT DEVELOPMENT

Plant development—the orderly outgrowth of organs and the progression from seed to seedling to vegetative plant and finally

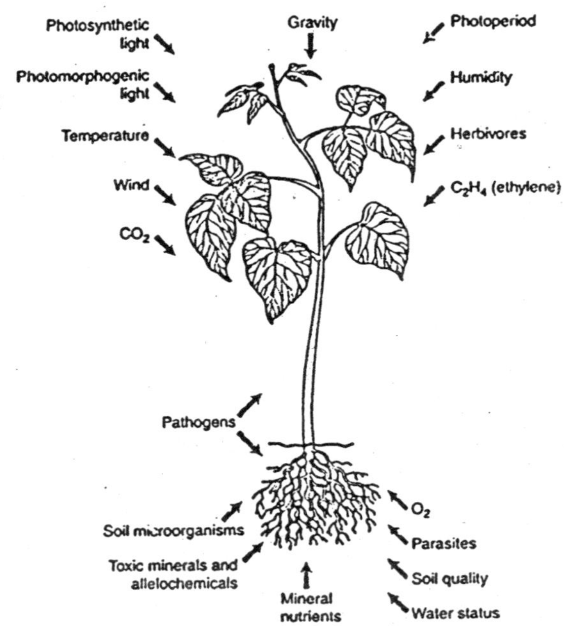

Figure 11.5: ***External signals that affect platn growth and development include many aspects of the plant's physical, chemical, and biological environments. Some external signals come from other plants. Apart from gravity, all other signals vary in intensity, often from minute to minute.***

to flowering plant—is profoundly affected by the environment. Indeed, a multitude of environmental stimuli or signals affect plant morphology and development. Many plant species heed environmental cues, such as changes in day length or a cold period, to

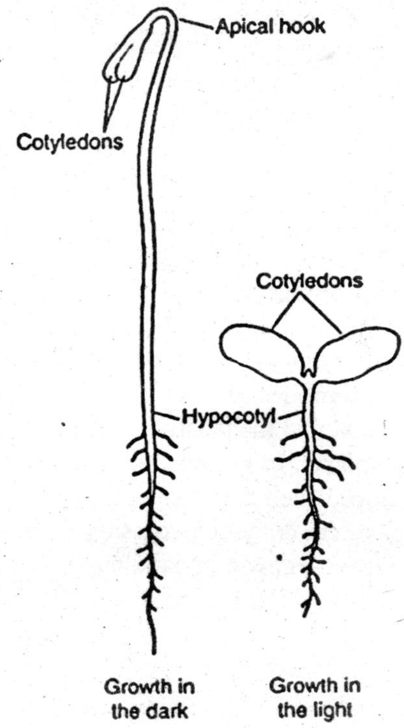

Figure 11.6: ***Comparison of dicot seedlings grown in the dark and in the light. Darkgrown seedlings are yellowish (etiolated) and tall, with a long hypocotyl (young stem) and a clearly defined apical hook and unexpanded unopened cotyledons. The darkgrown seedling is much shorter and dark green. The hook has straightened, and the cotyledons have expanded. The first true leaves are not yet visible at this stage.***

proceed from one developmental stage to the next. The amount of light received, spectral quality of that light, relative lengths of day and night, and temperature regime during day and night development. There is, first of all, a quantitative relationship both between light intensity and growth, and between temperature and growth. Up to a point, plants generally grow more vigorously if the temperature is higher and if there is more light. However, because plant have evolved in specific environments to which they are adapted, they may have quite differetn optimal light intensities and temperatures. Plants adapted to growing in dim light— on the floor of the tropical forest, for example—suffer from light stress and photo-oxidation if planted in full sun. Apart from these general effects of temperature and light, specific environmental regimes are often needed to let plants progress from one stage of development to the next.

Many species require light of break seed dormancy,and seeds buried too deeply in soil do not germinate. Moreover, seedlings need light to start greening. When seedlings emerge from soil, they are faintly yellow (etiolated); light causes them to turn green. In addition, when light touches the seedling the stem elongates less rapidly, and the internodes are much shorter. In dicots the leaves expand dramatically, and the plastids in leaf parenchyma cells develop into chloroplasts. In grasses, the leaves unroll and chloroplasts also develop. When the leaf parenchyma cells are exposed to light, they make the enzyme necessary to carry out photosynthesis; but if the leaves remain in darkness, they do not make these enzymes. If you think about this process in genetic terms, you see that light activates hundred of genes that encode all the proteins needed for leaf developments and photosynthetic function.

Reproductive growth or flowering can also depend on a particular environmental regime. As noted earlier, some biennial plants, including many varieties of winter wheat, absolutely require a period of cold weather before the vegetative apical meristem can become a reproductive meristem and give rise to flower primordia. This process is called *vernalization*. Winter wheats are normally planted in the fall and grow 10-20 cm tall before winter arrives. The wheat overwinters under a blanket of snow, and growth resumes in the spring. To come into flower and produce seed, winter wheat needs the cold period. Many other plants, including wild carrot, need to vernalize. The relative lengths of the day and the night determine whethers flowering will occur. This phenomenon, called *photoperiodism*. The photoperiod also has other effects of plant development. For example, the development of fleshy storage roots (such as carrots and beets) is governed by day length.

Scientists can mimic some of these environmental effects on plant development by applying plant hormones. During a developmental transition, the level of a particular hormone within the plant may change dramatically, and if the plant is sprayed with a solution of the same hormone, the transition occurs even in the absence of the correct environmental regime. For example, spinach normally requires long days and short nights to form flowers; scientists can circumvent this environmental requirement by spraying the plants with the hormone gibberellic acid. Analysis of unsprayed plants shows that levels of specific gibberellins rise as the plant make the transition from vegetative to reproductive growth. Such observations have led to the hypothesis that changes in the hormone level within the plant mediate the effects of the environment.

DEVELOPMENT OF SEEDS

The plant life cycle differs from the more familiar mammalian one in that male and female reproductive organs usually occur on the same plant and more than often than not in the same flower. The cells of flowers are diploid, with two sets of chromosomes. Some specialized cells in the reproductive organs undergo meiosis, a process that halves the chromosome number and produces haploid cell. These haploid cells undergo a few normal mitotic cell divisions, and some of their progeny then differentiate either into haploid egg cells or into haploid sperm cells. In animals, the haploid cells produced by meiosis differentiate directly into the sex cells (gametes) and there is no haploid developmental stage. The union of haploid egg and sperm produces a diploid zygote, which is the first cell of the new organism. Further cell divisions of this zygote result in the formation of the embryo, which forms a seed and then a new plant. Formation of the embryo and the seed occurs within the tissues (ovary) of the another plant.

Flowers and Fertilization. We said earlier that conversion of the vegetative meristem to a flower meristem initiates sexual

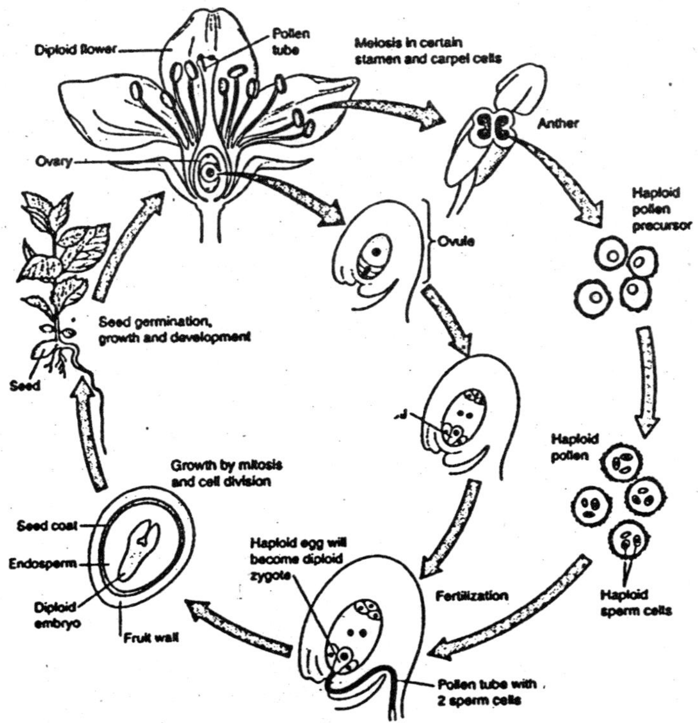

Figure 11.7: ***The life cycle of a plant showing the production of gametes (sex cells) and their union at fertilization. The flower, top left, is diploid and contains male (stamens) and female (carpels) reproductive organs. Specialized cells in each of these undergo meiosis to prodcue haploid egg and sperm cells. A single egg cell is formed in an ovule in the carpel. Two sperm cells are present in each pollen grain produced within the anther of the stamen. Fertilization results in the formation of a diploid zygote, which develops into an embryo as part of a seed. The seed germinates to form a vegetative plant, which eventually produces flowers.***

reproduction by forming reproductive organs. The flower organs occur in four concentric rings at the end of the stem: sepals, petals, stamens, and carpels (Figure 11.8). The individual carpels are usually fused to form a pistil. Sepals and petals often have the same shape at leaves, but stamens and carpels look quite different. The pistil has three parts: a swollen ovary with one or more ovules at the base of a style, and a stigma at the top. Pollen grains that land on the stigma germinate, and pollen tubes grow into the stigma and style to find the egg cell in the ovule. When the pollen tube reaches the ovule, it bursts and spills out two sperm cells that fuse, one with the egg cell and one with another cell that will produce the endosperm. This double fertilization requiring two sperm cell is unique to flowering plants.

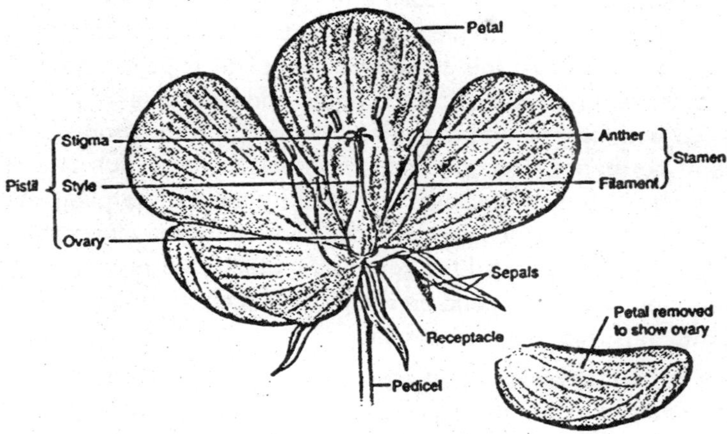

Figure 11.8: ***Flower structure. Most flowers have four types of organs: (1) sepals, often green, that protect the unopened flower, (2) petals, often coloured, that attract pollinators, (3) stamens, which produce pollen with sperm cells, and (4) a pistil consisting of fused carpels (individuals carpels are not shown). Carpels contain ovules and one cell in each ovule undergoes meiosis. Of the four daughter cells formed in meiosis, three die and only one forms the egg cell.***

Flowers structures vary from plant to plant, but in general the flowers of the important cereal crops occur in clusters. In wheat and rice, small flowers are connected to a central stalk termed a *spike*; here, both male and female reproductive organs are in the same flower. For maize, in contrast, female flowers are located in lateral ears and male flowers in terminal tassels, at different locations on the same plant.

The union of egg and sperm that initiates development of a new plant occurs within the ovule. Depending on the species, pollen may come from the same plant (self-pollination as in wheat) or from another plant of the same species (cross-pollination, as in maize). This type of fertilization that normally occurs for a particular species has important implications for plant breeding and for creating hybrid seed.

Embryogenesis. After fertilization, the zygote—the cell that results from the union of egg and sperm—undergoes many cell divisions and forms a multicellular embryo. The growth of the embryo occurs within the ovule in a rich nutrient medium called the *liquid endosperm*. At this stage, the new plant is growing completely heterotrophically, and the liquid endosperm provides the embryo with sugars, amino acids, vitamins, and the proper balance of hormones. The endosperm synthesizes these substances from sucrose and amino acids provided by the mother plant. As the embryo grows, it acquires more and more biosynthetic capacities; soon it can make all the complex molecules it needs, but it always continues to depend on a supply of sucrose, amino acids, and minerals from the mother plant.

As important aspect of embryo formation is the organization of the embryo axis: The two meristems, the SAM and the RAM, are located at opposite poles of the embryo. The asymmetric distribution of the hormone auxin plays a major role in this process. But that asymmetric distribution implies that all cells in a young embryo are not alike. So how soon do the cells become different from each other? Scientists have studied this process in greatest detail in *Arabidopsis thaliana*; at the 16-cell stage, the embryo is already divided into an upper half (the future SAM and cotyledons) and a lower half (the future RAM and stem of the seedling). By looking at the expession of specific genes at the 16-cell stage, researchers can find genes that are expressed only in some of the cells. Such observations suggest that cell differentiation in the embryo begin very early indeed.

Seed Formation. Once the embryo is formed, food reserves (proteins, oils, starch, and minerals) are synthesized and begin to accumulate in the storage parenchyma cells of the cotyledons or in other tissues closely associated with the embryo. In the majority of dicots, most of the reserves accumulate in the cotyledons, which themselves become very large. In cereals and other grasses, most of the food accumulates in the endosperm, a tissue located just next to the embryo. Food reserves accumulate in specialized organelles within the cytoplasm: starch in amyloplasts, a special type of plastid; protein in small protein storage vacuoles (Figure 11.9); and oil in very small oil bodies.

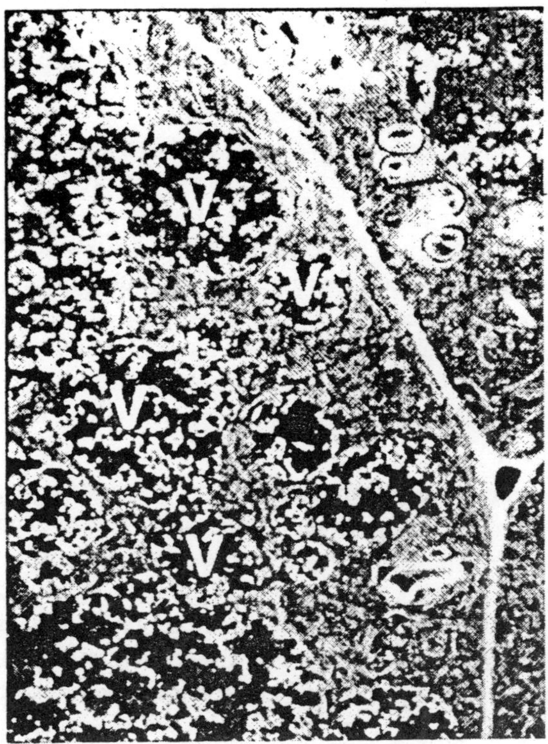

Figure 11.9: ***Legumes seeds are a rich source of protein stored in vacuoles. In this electron micrograph of a soybean cell, the vacuoles (V) are not yet full and protein will continue to accumulate during the 40-day seed maturation period.***

During the development period that stretches from fertilized egg to mature seed, three important genetic programs are activated: The first regulates cell division and the formation of tissues and organs in the embryo; the second regulates expansion of the storage organs and the massive biosynthetic effort to make reserves.

Finally, the seed needs to prepare for survival in the dry state and for responding to environmental cues that bring it out of dormancy. Survival in the dry state (15% water) depends on accumulating sugars, oligosaccharides, and special proteins that bind water and protect the cytoplasm components from damage when the cells dry out.

FORMATION OF THE VEGETATIVE BODY

With the appearance of conditions that break seed dormancy and allow germination, the seed restores its former water content and swells. The rehydration triggers the two main events of germination: embryo growth and use to the seed's food reserves. Embryo growth depends on using the reserves, because not until green leaves have formed can the seedling grow autotrophically. Its initial growth depend on the carbohydrates, fats, proteins, and minerals in the storage tissues of the seed. The growing embryo first breaks down the starch, fats, and proteins into their simple constituents (sugar, fatty acids, and amino acids), and then absorbs and uses them.

As the embryo absorbs the digested products of the seed's storage material, it uses them to build more cells and to expand cells once made. Within a few days, a recognizable root is growing down through the soil and the shoot (seedling stem and cotyledons) pushes above ground. Once above ground, light causes the stem to elongates less rapidly, and the leaves to expand. In the light, chlorophyll is synthesized and the plant begins to carry out photosynthesis and ceases to depend on seed reserves.

Formation of the Root system. We noted earlier that cell growth involves three basic processes: division, enlargement, and differentiation. In the root tip these are spatially separated. At the very tip, just behind the protective root cap, is the RAM, a region of *continuous cell division*. These new cells become about 100 times larger; behind the region of cell division there is a region of *cell elongation*. Finally, as they are enlarging and after they have reached their full size, the cells specialize, depending on location, into the various cell types of the root; behind the enlargement region there is thus a region of *cell differentiation*. As the root grows through the soil, this sequence is always maintained.

Root growth is accompanied by the formation of lateral roots, which in turn form other lateral roots, until the plant has established a widely branched root system capable of taking,up water and minerals from a large volume of soil. The size and shape of that root system are genetically determined, and some plants form very deep root systems, others very shallow ones. Like many other feature of the plant, these represent ecological adaptations. There may be considerable variation in the root systems of wild relatives of crops that might be exploited by plant breeders. However, unlike shoot systems or flowering characteristics, root systems are difficult to study because people cannot see them. Moreover, given the differential growth of the roots in different soil types and under different climatic conditions, traditional plant breeding for better root systems is almost impossible. Nevertheless, molecular geneticists using genomic information will be able to determine in the near future the important genes that determine the characteristics or root systems. What now seems impossible to achieve with traditional breeding may be within reach the tools of molecular breeding.

In addition to having an extensive root system, many plants greatly growth increase the area exposed to the soil by forming root hairs. These are tiny projections from individuals epidermal cells. Roots also form symbiotic associations with soil fungi called *mycorrhizae*. These serve the same purpose as root hairs: The fungal threads take up minerals from the soil and transfer them to the root cells for transport into the plant. To provide the plant with water and mineral nutrients, the root system must grow continuously, and new root hairs are always being formed as the older ones die off.

Storage of food reserves is another important function of the root system. Some biennial plants, such as beets and carrots, develop a thick, fleshy taproot at the end of the first growing season. The reserves stored in this taproot are used at the beginning at the second growing season, and a new shoot produces flowers and seeds. Other plants, such as cassava, develop numerous fleshy, tuberous roots for food storage and vegetative reproduction.

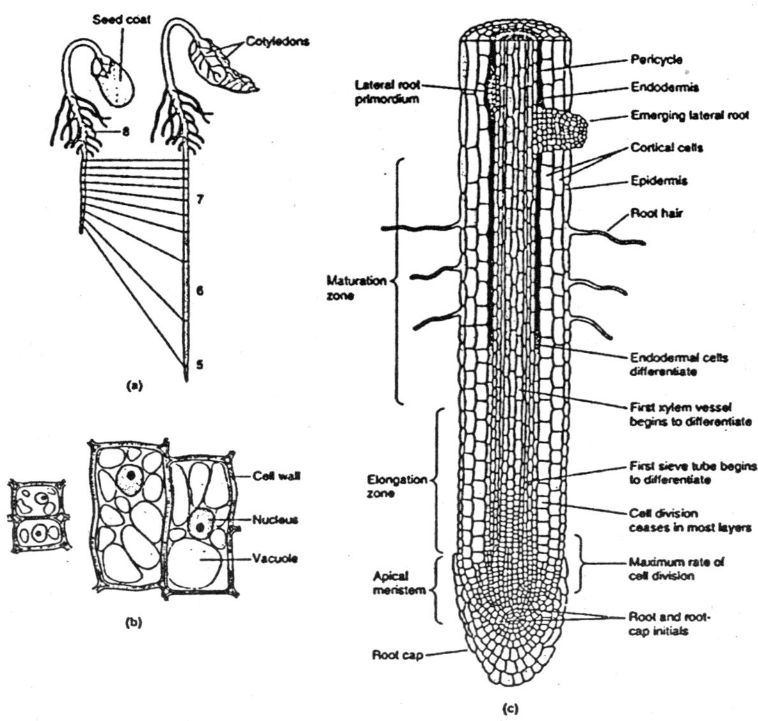

Figure 11.10: *Diagram showing root growth. (A) The distribution of growth zones in the root: 5 = apical meristem, 6 = zone of rapid elongation, 7 = elongation has ceased, 8 = mature portion; (b) cells are shown enlarged about 1,000 times; small cells have very small vacuoles and enlargement means a big increase in vacuolar volume; (c) detailed view of the different zones on the root: root cap, root apical meristem, elongation zone, maturation zone, and lateral root.*

Formation of the Shoot. The processes involved in shoot growth are generally similar to those for root growth except that the three basic proceses are not spatially separated. As noted earlier, a shoot consists of a stem and of leaves attached to the stem at points called *nodes*. Of particular interest to plant physiologists and plant breeders are the processes that control stem elongation in the shoot. Scientists know that at the cellular level hormones influence elongation. Auxin, gibberellins, and brassinolides all affect stem elongation in different plant species and under different conditions. Cell elongation requires, among other processes, that the network of large molecules in the cell wall loosen; this loosening depends on a slight increase in acidity and on enzymes and other proteins secreted from cells into the wall. Hormones activate some of the genes that encodes these enzymes.

Dwarf mutants with short stems in which the cells fail to elongate have been found for a number of plant species. Such mutants either have very low levels of a particular hormone (such as gibberellin) or are unreponsive to the hormone they synthesize. As part of the development of high-yielding wheat varieties for the Green Revolution, breeder purposefully selected dwarf because shorter plants with large heads of grain do not fall over (lodged) so easily. Moreover, if the plant doesn't waste energy making stem tissue, it can channel more energy into seeds.

Vegetative Reproduction and Propagation. In addition to sexual reproduction marked by seed formation, many plants can also reproduce asexually. In asexual (vegetative) reproduction, new plants form, not from seeds, but from specialized structures of the root, stem, or leaf (the vegetative organs of the plant). In nature, such vegetative reproduction often proceeds by means of horizontal stems or roots, which let the plant spread over a larger area. For example, the strawberry plant sends out runners (horizontal stems), which make new plants by sending down roots and forming leaf clusters where they touch the ground. In some plants, horizontal stems that grow underground may become thick and flaehy and form tubers, as in white potatoes. Each tuber can give rise to a new plant. People have learned to encourage the growth of such horizontal stems by mounding soil around the plant to cover the lower portion of the shoot. Tubers formation is also controlled by day length and mediated by genes that direct the growth of these structures and control starch accumulation. In other plants as well, the specilalized structures involved in vegetative propagation are at the same time food storage organs and enable the plant to survive adverse conditions.

Humans use of the natural process of vegetative reproduction is termed *vegetative propagation* or *cloning*. People use cloning to propagate potatoes, sweet potatoes, sugar cane, berries, nuts, and a variety of fruit trees and ornamental plants. The advantage of asexual reproduction is that all the offspring have exactly the same characteristics (genetic makeup) as the parent plant. This homogeneity is especially important when the parent is a genetic hybrid whose sexually produced offspring would not be the same as the parents. For example, an apple tree grown from a seed does not usually produce apples of the same quality as the parent plant. Vegetative propagation through grafting ensures uniform quality in such plants.

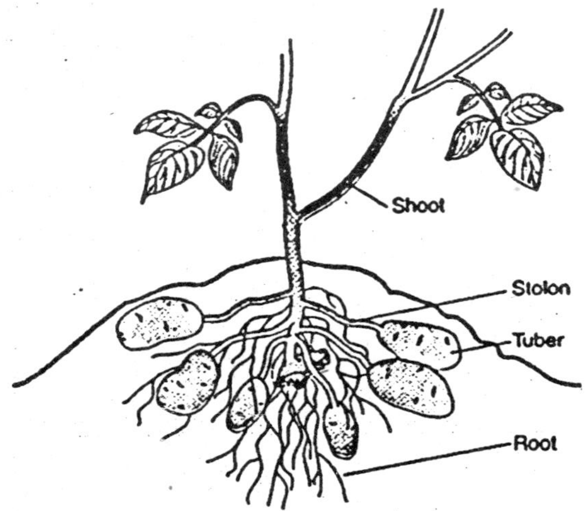

Figure 11.11: ***Vegetative reproduction of potato (Solanum tuberosum). Potato tubers are attached to underground horizontal stems or stolons that are attached to the main stem.***

Plants that produce specialized structures for vegetative reproduction are easily propagated. Thus pieces of white potato tubers or of the tuberous sweet potato roots sprout and proudce new plants when they are put in the soil. Other plants, such as sugar cane, pineapple, cassava, and many ornamental plants, are propagated by stem cuttings. In these species, pieces of stem produce roots spontaneously when placed in moist soil. The discovery that in many species the plant hormone auxin promotes the rooting of stem cuttings has greatly enlarged the list of plants that can be propagated this way. Many stem cuttings that would not ordinarily produce roots do so after the lower end of the stem is dipped in a solution of auxin.It is also possible to dissect out the meristem of a plant, cut it into small pieces, and regenerate new plants. This procedure, called *meristem culture*, must be done under sterile conditions. The goal of meristem culture can be genetic uniformity of superior strains, as with oil palms or orchids, or the rapid propagation of plants free of pathogenic viruses, as with strawberries.

REPRODUCTIVE DEVELOPMENT OF FLOWERS

We noted earlier that sex cells or gametes are produced in the reproductive organs of the flowers. Most dicot flowers consist of four concentric circles of organs. The sepals, which are often small and green, completely enclose the flower bud before the flowers opens. Next are the petals, which are often colored and showy, especially in plants pollinated by insects and birds. Inside the ring of petals are the stamens, and at the center is the pistil. Some of the most important crops—wheat, rice, maize, barley, and millet—are of the grass family (Poaceae), in which the flowers have a somewhat different structure. The inner two concentric rings of floral primordia also develop into female and male reproductive organs, but how the other organs of the flower correspond to sepals and petals is not clear. Grass flowers are called *spikelets*, and each flower has two leaflike organs, the lemma and the palea, that enclose the male stamens and female carpel. In wheat, barley, oats, and other grains, a long, slender awn is attached to the palea. When these flowers are mature, the lemma and palea separate slightly, exposing two stigmas and allowing the anthers to hang free. These wind-pollinated flowers have no need of showy petals to attract pollinators.

The elaborate floral organs, each with their own tissues and cell types, originate in the flower bud from floral primordia, just like leaves originate as leaf primordia. The development of these floral primordia is itself precisely regulated by genes, and many variations to the basic flower structure result from variations in the development pattern. For example, some flowers contain only male or only female sex organs—as in maize—because one set of organs fails to develop in either the male or the female flower. Usually both types of flowers are present on the same plant, but in some species each individual is genetically either male or female.

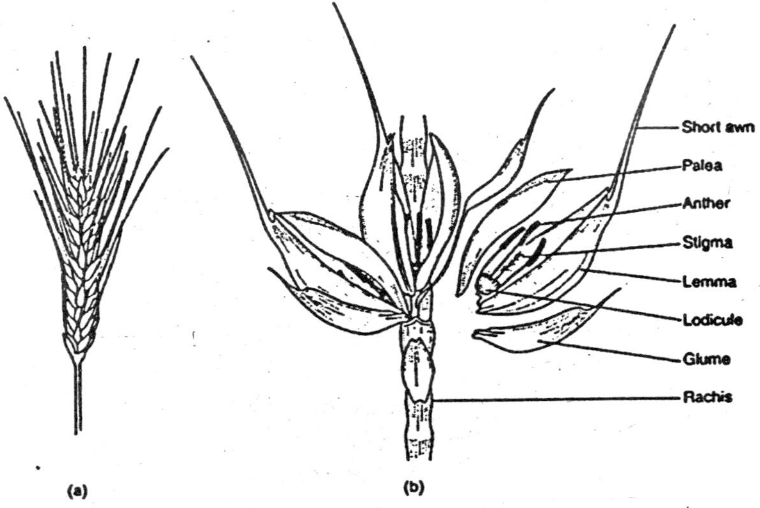

Figure 11.12: ***Structure of a grass flowers. (a) Spike of wheat (Triticum aestivum). (b) Spikelets attached and detached from the rachis of wheat. The glume is thought to correspond to a bracht, a small leaflike structure found on many plants. The lemma and palea are sepal-like organs, based on the types of genes expressed there. The lodicule, at the base of the lemma and palea, swells up and causes the flower to open. The genes expressed in the lodicule indentity it as equivalent to the petals of a dicot flower.***

In many plants, a flowering signal or stimulus that originates in the leaves can convert a terminal vegetative bud that gives rise to leaves and stem tissues into a flowering bud. The chemical identity of this signal remains undiscovered. Once the SAM has received the flowering stimulus, the SAM changes into a floral meristem and starts forming floral organ primordia that differentiate into floral organs. These processes are controlled by gene cascades in which the protein product of one gene activates the next gene.

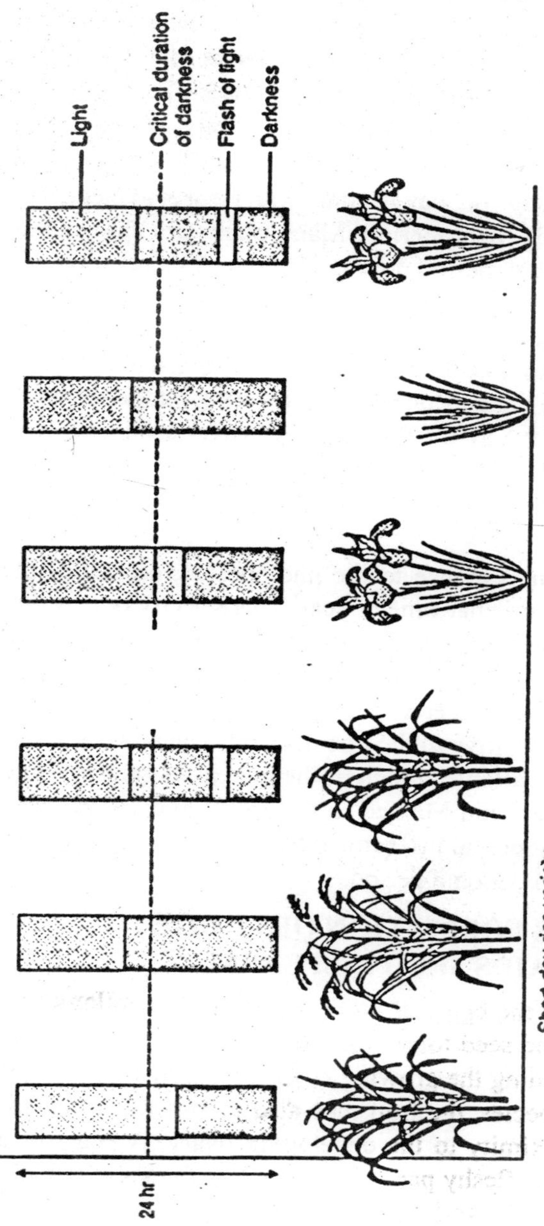

Figure 11.13: *Photoperiod regulation of flowering. The photoperiod is the number of hours of light and dark within a 24hr period. Short-day plants (left) flowers when the length of the night exceeds a critical number of hours, if the night is long but interrupted by a few minutes of light, they do not flowe, shown here is a short-day rice variety. (Many varieties of rice are day-neutral plants). Long-day plants flower when the night falls below a critical number of hours (right), if the night is long but interrupted, they flower.*

A combination of genes controls the identity of each organ. Reproductive development is often extremely sensitive to stress, and a single day of heat stress at the wrong time is often enough to block one or more aspects of this important phase of development.

As mentioned, when it comes to initiating flowers, many plant species respond to the photoperiod. Plants can be classified as "short day", "long day" or "day neutral" with respect to their flowering response to the photoperiod. Many plants are day neutral, meaning that the photoperiod does not influence them. Although they are called "short day" plants and "long day" plants, they are really long night plants or short night plants, respectively. Indeed, plants do not monitor day length, but rather night length. You can easily demonstrate the dependence of flowering on the length of the night (or the day) by turning on a light over such a plant in the middle of the night for a few minutes. This short light period alters the plant's respond, even if the length of the day remins the same.

Because of their response to the photoperiod, certain plants flowers only at a particular time of year. This selectivity may disadvantage the farmer, who thus can grow only one crop per year. Multiple cropping—which farmers practice in many tropical areas that have abundant water and sunlight—is only possible if the crop plants are insensitive to the photoperiod. The new strains of wheat and rice that allowed much of the increased grain production in India, China, and Mexico in the 1970s and 1980s, were selected to be photoperiod insensitive, so that farmers could grow several crops in a 12-month period.

THE FORMATION OF FRUITS AIDS SEED DISPERSAL

Fertilization of the egg cell be the sperm cell is followed by embryo growth and seed formation. Along with seed formation, the ovary surrounding the growing embryo(s) forms a fruit. Depending on the species, the fruit may also incorporate other tissues in close proximity to the developing ovary. A peach is a "typcial" fruit. The fleshy part of the peach is the greatly thickened wall of the ovary, and the seed is inside the stone.

The explanation just given should make clear that any structure containing seeds is a fruit, not a vegetable. Tomatoes, peppers, eggplants, zucchinis, cucumbers, okra, and green beans are all fruits. Each has a more or less thick, fleshy wall that originates in the ovary and contains seeds. So why do people call them *vegetables*? Well, the word *fruit* comes from the Latin *frui*, meaning to enjoy, to delight in, to have the use of something desirable. Sweetness in innately the most preferred taste sensation of mammals, and peoples call *fruits* those parts that are fleshy, sweet, and enjoyable. (So why is a lemon a "fruit"? Language and customs can be mysterious!)

The early development of the fruit-also called *fruit set* of *setting*—depends on hormones, especially auxin and gibberellin, produced by the growing embryo within the seed. These hormones stimulates the cells of the ovary to divide and expand. Spraying with synthetic auxin is sometimes used to promote fruit set and cause the development of fruits without fertilized seeds. In addition, plant breeders have found mutants in which fruit development does not depend on the hormones produced by the seeds. Such mutants often produce seedless fruits. The fruit is the vehicle that aids seed distribution. Dandelion fruits (commonly called *seeds*, but to a botanist they are fruits) are carried away (for example) by the wind, berries are eaten by small mammals whose droppings are scattered over the countryside, and acorns are carried away and buried by rodents. The seed pods of many plants split open quite violently when they are completely dry, scattering the seeds in the process. This property, called *dehiscence* (or shattering), has disappeared from cultivated crops, which retain their seeds rather tightly.

The later developments of the fruit, also called *fruit ripening*, involves a number of changes that usually soften the fruit wall and convert acids and starch to sugar. At the same time, the fruit synthesizes aromatic substances to attract animals. In many species the ripening process is controlled by ethylene, which triggers an entire genetic program when the fruit produces it. Commercial fruit companies often pick fruits green, before they produce ehylene, and then reduce ripening by keeping the fruits for a few days in storage rooms containing ethylene or another gas that has the same activity. This procedure lets people handle and transport the fruit before they soften as a result of natural ripening. However, the flavour of such fruit is never as good as that left to ripen on the plants.

INTEGRAL PROCESSES OF DEVELOPMENT

The death of individual cells, groups of cells, or entire organs is an integral part of development. The best known example of this phenomenon is the shedding of leaves by trees. Whether on deciduous trees or not, after leavers mature they gradually start losing their capacity for photosynthesis and undergo senescence. Such senescence is visible at the ultrastructural level: The internal membranes of the chloroplast become disorganized, and little fat globules accumulate in the chloroplasts. Eventually the cells undergo the process of autophagy (self-eating) in which the major macromolecular components (proteins, nucleic acids, lipids and starch) are digested and the resulting small molecules are exported to the growing part of the plant or to storage tissues. Thus, when a tree sheds yellowed leaves, these are not full of protein and minerals—everything has been digested and exported. In the final phase, the cells digest themselves and only the cell walls remain. Because the organism controls both the initiation and the execution of this senescence, the process is called *programmed cell death*. Programmed cell death may be triggered by environmental factors (annual leaf shedding is a photoperiodic response in many plants) and is controlled internally by hormones. Once again, the expression of specific genes underlies the whole process.

When you look at different aspects of plant development, you can see that programmed cell death occurs all through the plant and throughout its life. To begin with, digestion of reserves stored in the cotyledons of many dicots, or in the endosperm of cereals, is followed by cell death. In the aleurone cells of cereals, which secrete enzymes that digest stored reserves, cell death is accelerated by gibberellin and retarded by abscisic acid. As vascular tissues develop, precursors to vessel elements undergo programmed cell death and these elements do not become fully functional until the cells die. The sclereids in ground tissues also undergo programmed cell death. Another example of programmed cell death is senescence of sepals, petals, and stamens after fertilization.

When the roots systems of some plants, such as maize, are flooded and thereby deprived of oxygen, they can develop internal channels through programmed cell death. Water has much less oxygen than does the soil atmosphere, and lack of oxygen triggers this developmental process. The channels permit faster diffusion of oxygen throughout root tissues. The channels develop in the cortex, and the cortex with its channels is now called the *aerenchym*. The channels form when the cells are triggered to undergo programmed cell death and enzymes synthesized as part of aerenchyma formation digest the cell walls. This entire process is regulated by ethylene, the synthesis of which is induced by the absence of oxygen.

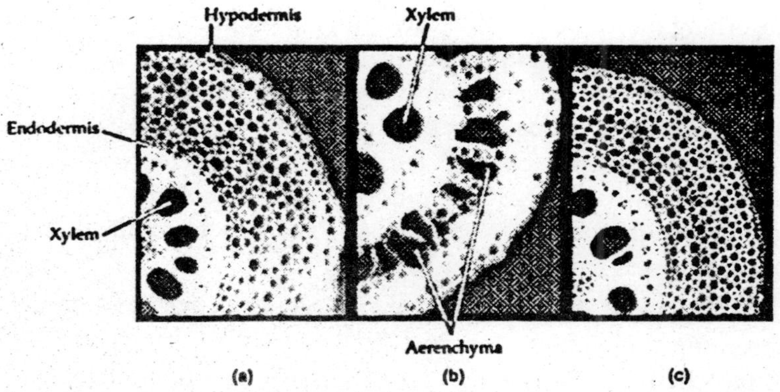

Figure 11.14: ***Aerenchyma formation in maize roots. (a) Cross section of control root of 11-day-old plant. (b) Plant exposed to an unaerated nutrient solution for 4 days, depriving the roots of oxygen. (c) inclusion of 50 micromolar EHTA in the nutrient solution. EGTA penetrates the root cells and binds up calcium, making it impossible for calcium to fulfill its signaling function.***

The cellular program leading to programmed cell death involves the fragmentation of nuclear DNA, caused by enzymes called *nucleases*; the biosynthesis of enzymes that digest proteins, lipids, and RNA; and the vacuole's engulfing of cytoplasmic structures that vacuolar enzymes then digest. The genes that encode all these enzymes are activated as part of the programmed cell death phase. Scientists have known for some time that hormones are involved in different aspects of senescence, but they are only now exploring the genetic basis of programmed cell death.

HORMONAL SIGNALS TO THE NUCLEUS TO REGULATE GENE ACTIVITY

The preceding description makes clear that hormones and the environmental regulate many aspects of plant development. Because development requires the orderly expression of genes in specific genetic programs, the question arises, How does the genome "know" when to activate the correct genes? How is the environmental signal transduced to the genome? How does light turn genes on? How does ethylene turn on the genes that encode the enzyme needed for fruit ripening?

To perceive a light signal, there must be a receptor, a light-absorbing molecule that interacts with a protein to start the chain of events that leads to gene activation. To perceive a hormonal signal, there must be a receptor to which the hormone binds. The interaction of the signal (whether light or a hormone) with the receptor usually subtly changes the shape of the receptor, affecting its activity or its ability to interact with another protein. This protein in turn passes the "signal" on to the another protein, and so on, until genes in the nucleus activate.

In the case of light, the receptors are pigments that have a light-absorbing part and a protein part. Such pigments belong to a class of protein called *chromoprotein*. (Hemoglobin is a typical chromoprotein). Plants are green because they contain chlorophyll, but in addition to chlorophyll they contain blue chromoproteins, called *phytochromes*, that specifically absorb red light, and they contain chromoprotein, which are yellow chromoproteins that absorb blue light. Both phytochromes and cryptochromes are present in very small quantities, but these amounts are nevertheless large enough to register subtle changes in the light environment and to affet plant development. Plant use these pigments to measure the presence and absence of light, duration of the dark period (photoperiod), intensity of light, and spectral equality of light (in the shade underneath a tree canopy, there is a relatively more red light).

The absorption of light energy by the light-absorbing portion of the molecule changes the shape of the protein. This change causes phytochrome to acquire a negatively charged phosphate group, which now permits phytochrome to enter the nucleus through pores in the nuclear envelope. It may well do so bound to other proteins. It has recently been shown that phytochrome's entry into the nucleus is followed by its binding to proteins responsible for DNA transcription. These DNA transcription complexes then bind to the promoters of genes whose activity is up- or down-regulated by the light signal.

Investigating signaling pathways is now a very active research area, not only to understand how they work, but because scientists may be able to alter the course of plant developmental processes if they interferes with the signaling pathways. For example, how does the plant perceive the gaseous hormone ethylene? In this case the receptor is a protein in the plasma membrane that has a domain (a portion of the protein) on each side of the membrane, with a short piece of the polypeptide spanning the membrane. Ethylene binds to this receptor, and binding turns the developmental switch that precipitates a phosphorylation cascade, meaning that phosphate groups become attached to a protein and are then transferred to the next protein, and on to the next. Many organisms use such cascades to sense what is going on in the environment and to respond. Remarkable, yeast cells use a very similar system to measure salt concentration in their environment. Eventually such a cascade ends with a change in activity of a number of transcription complexes, and genes are activated or silenced.

In many cases, the signal causes a rapid influx of calcium into a localized area of the cytoplasm and the calcium is equally rapidly brought back to its basal level. This is referred to as a *calcium spike*. The influx of calcium occurs because proteins that form calcium channels in certain membranes (for example, the plasma membrance or the tonoplast) are opened. Other proteins, called *calcium pumps*, in those same membranes bring the level of calcium down again. However, this brief elevation of cytoplasmic calcium permits certain enzymes that requires câlcium to be active for a short while. This activity propagates the signal in the cell. The ethylene-induced formation of aerenchyma is inhibited by EGTA, a substance that binds cytoplasmic calcium and does not permit the calcium spike to occur. Such experiments suggest that a calcium signal mediates this particular effect of ethylene.

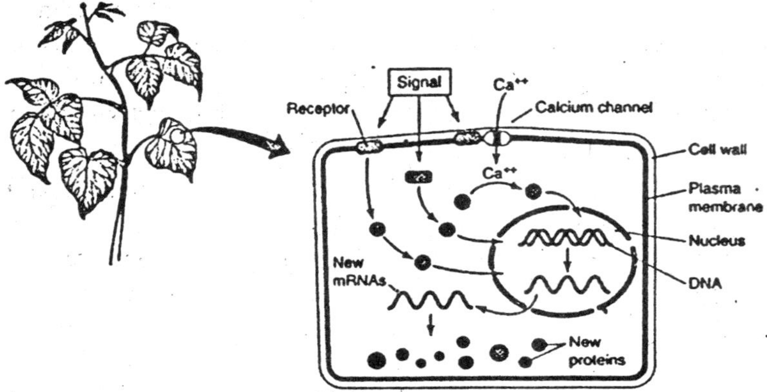

Figure 11.15: ***Signal transduction in a plant cell. A signal (light or a hormone or other molecule) reacts with a receptor (R), which is in the plasma membrane (PM) or in the cytoplasm. In one case, proximity to a calcium channel causes the channel to open, resulting in sudden influx of calcium. In each case, after proteins are modified down a "chain of command" that reaches the nucleus where genes are activated. New mRNAs are made and exported from the nucleus. In the cytoplasm, they direct the synthesis of new proteins.***

GENETIC MODIFICATION OF CROPS

Because genes and their expression are at the basis of all developmental processes—from fertilization to senescence—it stands to reason that scientists can alter development by introducing new genes or by changing the expression of existing ones. Actually, many crop plants differ from their wild relatives by having bigger seed or tubers, because specific developmental processes have been eleminated or altered. The modification of their developmental makes them more suitable as crops. Few crops would actually survive in the "natural" world, as their development has been too drastically modified. As a result, crops generally do not escape and become weeds. The following examples

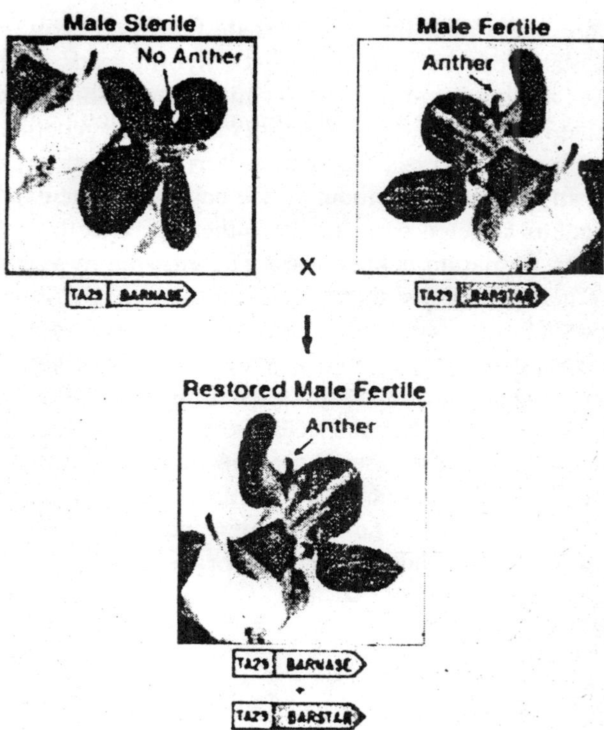

Figure 11.16: ***Genetics engineering for male fertility control in canola. Canola was tranformed with a gene that encodes a ribonuclease enzyme (barnase) or a ribonuclease inhibitor (barstar). In each case, the regulatory region (promoter) used to drive gene expression was TA29, which directs gene expression specifically to the tapetal layer in the anther. In the plants expressing barnase, the flowers have no anters and they are male sterile. In the plants expressing barstar and in the barnase × barstar hybrid, the flowers have normal anthers.***

show how some ways in which gene technologies are modifying crop developments.

Engineering Male Sterility and Hybrid Seed Production. Hybrid plants produced by crossing two different line often show increased yield because of hybrid vigor. However, large-scale and cost-effective production of hybrids requires the availability of male sterile lines and a way to restore made fertility in the next generation. For some crops, there are natural male sterile mutants and other mutant lines that restore male fertility. A combination of such plants can be used to produce hybrid seeds, if the two properties are bred into high-performing lines. This is how hybrid maize is produced. Male sterility can also be genetically engineered by the expression of a bacterial gene that encodes barnase, an enzyme with ribonuclease activity. If the expression of the gene is targeted specifically to the tapetal cells of the anther, pollen formation is prevented. The enzyme hydrolyzes the RNA in the tapetal cells, thereby killing them. Male fertility can be restored if in the same cells a protein is expressed that binds ribonuclease extremely tightly, so that there is no chance it will have any activity. Such a ribonuclease inhibitor, called *barstar*, is found in the same bacterial species from which the barnase gene was obtained. Scientists uses these two genes to create lines of canola that can be crossed to efficiently generate hybrid canola seeds. The hybrid seeds, now used in Canada, can raise canola yields by 20% above those from nonhybrid seeds.

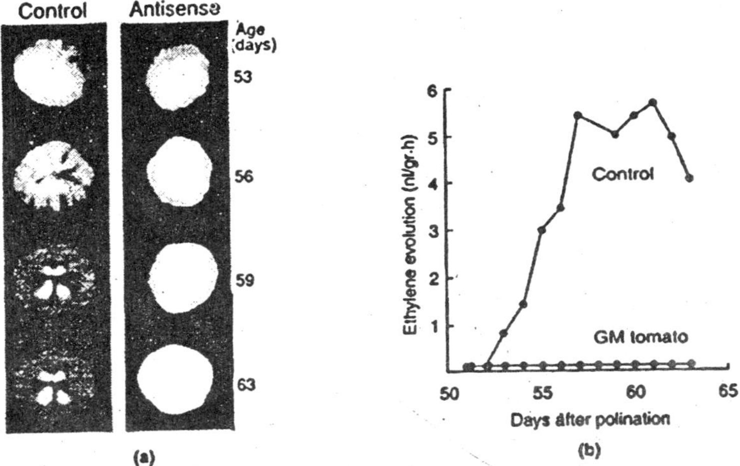

Figure 11.17: ***Genetic engineering for ripening control in tomato. Ethylene production in tomato has been reduced by genetically modifying expression of the gene that encodes one of the enzymes responsible for the ethylene biosynthesis. (a) The control tomatoes ripen normally between 53 and 59 days. The genetically modified (GM) tomatoes ripen extremely slowly during this same period. (b) Graph showing ethylene production in control and GM tomatoes.***

Modification of Fruit Ripening. We said earlier that ethylene regulates fruit ripening of many fleshy fruits and that the process can be speeded up by exposing fruit to ehylene. Ripening can also be slowed down by interfering with ethylene production or with ethylene perception by the cells of the fruit. Biosynthesizing ethylene requires two enzymes, and eliminating either one, by supporting the expression of these genes, results in fruits that make very little ethylene. Such fruits ripen very slowly, an advantage to farmers and customers: Such fruits can be left on the plant longer—it is no longer necessary to pick them while they are green—and can often be sold in markets that are farther away.

Modification of senescences with an Autoregulated Gene for Cytokinin Biosynthesis. In annual plants, seed formation is accompanied by senescence of the vegetative body, and nitrogen and other minerals stored in the leaves are used to make the seeds. So what would happen if the plant stayed green longer? Would it make bigger seeds, because it retains the capacity for photosynthesis? Plants such as lettuce and broccoli generally begin to senesce as soon as they are harvested. Could their senescence be delayed? Scientists had the idea of hooking the gene that encodes an enzyme for cytokinin biosynthesis to a promoter that is induced when cytokinin levels fall. Thus, as soon as the leaf begins to senesce, the gene for cytokinin synthesis is induece, and cytokinin synthesis accelerates, delaying senescence. The technology is being applied in Taiwan, with the goal of delaying senescence in broccoli after harvest.

Delaying seed pod splitting to avoid seed scatter. Most plants have evolved mechanisms to help them scatter seeds over a wide area. When the seed pods of a number of plants dry out, they split open quite abruptly—even explosively—scattering the seeds. This process, dehiscence, has disappeard from crops such as beans and peas because plant domestication selected against this trait. Dehiscence remains a serious problem in canola, where up to 20% of the seeds can be lost during harvest. Seeds scattered in this way germinate in the field the next year and pose a problem for the farmer who must now treat them as weeds. Scientists identified the genes responsible for dehiscence in *Arabidopsis*, a close relative of canola, and were able to generate mutant plants in which the pods do not split open. Commercial application of this discovery to canola is now being considered.

Figure 11.17: *Genetic engineering of flowering time in citrus. Flowering in citrus and other trees does not occur until the trees are several years old. The six-months-old plants on the right, which are flowering, have been transformed with a meristem identity gene. Such genes determine when a vegetative meristem converts to a floral meristem. The control plant on the left is not yet flowering.*

Genens that accelerate flowering. As noted earlier, flower formation depends on conversion of a vegetative meristem to a flowering meristem, a process controlled by the environment and by hormones that activate specific genes. Overexpression of such genes converts the vegetative meristem to a floral meristem much earlier in the plant's life than is normal. One application of this modification is in tree breeding. Many trees don't flower until they are 3 to 10 years old, making tree breeding a rather slow process. Tree breeding could be accelerated if the trees flowered after a year of growth.

These examples show that, in the future, genetic engineering of plants will move beyond introducing a single new characteristic with an unusual gene—as in the case of herbicide resistance—but will modify the development of the plant in subtle ways for the benefit of farming and food production. The new science of genomics will greatly help us identify the important genes.

12

Seeds Technology in Agriculture

Seeds play a central role in agriculture, serving as the means to propagate plants from one generation to the next as human food, as animal feed, and as an important commodity in international trade. Both people and domesticated animals ultimately rely on plants as their food source, which means both depend on seeds. Seeds are a concentrated source of carbohydrates, proteins, and fats for human and livestock alike, and are a significant source of minerals, vitamins, and fiber. They also provide raw material for industrial products. Cereals are the main source of food, followed by legumes and oilseeds, with soybean being both as legume and an oilseed crop (Table 12.1).

TABLE 12.1 : WORLD PRODUCTION (IN MILLION OF TONS) OF 15 MAJOR SEEDS CROPS, 2001

Cereals Crops	Production	Legume Seed Crop	Production	Other Oil seed Crops	Production
Wheat	518	Soybean	176	Rape /canola	36

Cereals Crops	Production	Legume Seed Crop	Production	Other Oil seed Crops	Production
Maize	605	Peanut	31	Sunflower	21
Rice	590	Common bean	17		
Barley	139	Dry peas	12		
Sorghum	58	Chickpea	8		
Oats	27				
Millet	29				
Rye	22				

Figure 12.1: ***Seed storage in Africa. Because seeds are dry, unlike roots and tubers, they can be stored for long periods of time. Nevertheless, they must be protected from predators and pests, especially weevils, whose larvae burrow into the seeds and eat the reserves. African farmers build elaborate granaries to rpotect their harvest.***

Seeds are important for other reasons. Most weeds sprout from seeds buried in the soil, and new weeds sprout each time the soil is cultivated. Because seeds are dry (6-15% water), they can be easily stored for long periods of time and shipped in vast quantities across the oceans. As propagules, seeds are the means by which genetically improved crops, whether achieved by conventional breeding or as a result of biotechnology, can be mass produced for use by farmers. Agricultural biotechnology companies purchased seed companies in 1990s, because genes specifying valuable traits can be sold to farmers as seeds. Finally, seeds are the means by which the genetic diversity in domesticated and wild crop species can be conserved in specialized facilities called *gene banks*.

The seeds of today are the products of thousands of years of domestication and decades of controlled breeding. All genetic improvements are passed from one generation to the next via the seeds. Domestication eliminated certain characteristics that benefit the survival of the species but are detrimental to the farmer. Elimination of seed dispersal mechanisms and seed dormancy occured quite early during domestication. Other selected traits were an increase in seed size, with its accompanying increase in stored reserves, and the production of more seeds per flowering head (inflorescence). For example, teosinte, the plant that is the closest living relative of maize and that probably resembles its wild progenitor, yields about 100 kg of seed per hectare. In contrast, modern maize hybrids can produce 8,000 kg of seed per hectare.

NUTRITIONAL RESERVES TO SUPPORT THE GROWING SEEDLING

Pollination followed by the fusion of a haploid sperm cell and a haploid egg cell, creates a diploid zygote. This process initiates seed formation. To understand the process, it is useful to first look at the end result, the mature seed. All seeds contain three essential parts: an embryonic root (radicle) and shoot (plumule), food reserve tissues or organs (cotyledons or endosperm), and a covering structure, the seed coat (testa). After germination, food stored in seeds supports seedling establishment until the plant becomes self-sufficient through photosynthesis. Because of selection and breeding, seeds of food crop species now contain far more reserves than are required for seedling growth. The mature seeds of cereals and legumes are distinctly different in structure. In cereals, most food reserves are in the endosperm, a tissue that is not part of the embryo, whereas in legumes such as soybean or the common bean, reserves are in the seed leaves or cotyledons, which

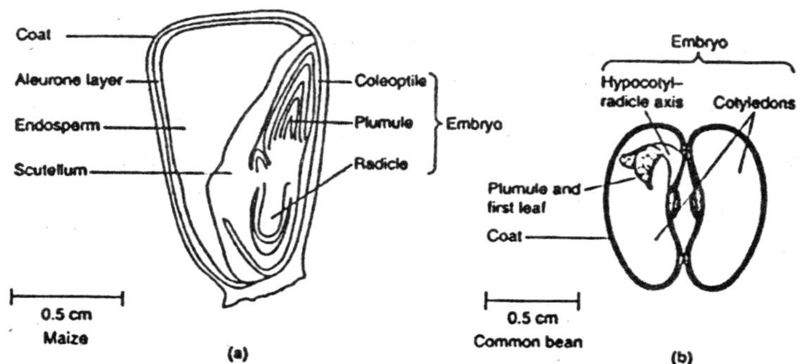

Figure 12.2: ***The structure of a cereal and a legume seed, using maize (Zea mays) and common bean (Phaseolus vulgaris) as examples. (a) In maize, the major site of stored reserves is in the endosperm, which is nonliving at maturity, except for the peripheral aleurone layer. (b) In bean, the cotyledons are the site of stored reserves, and the cells remain alive at maturity.***

Seed development can be divided into three major phases, each with its own genetic program. In the first phase, all the structures and tissues of the embryo form as a result of cell division and cell differentiation; in the second phase, the seed grows larger and the genetic program for synthesizing seed reserves is activated. Finally, during seed maturation the seed dries out and prepares to survive in the dry state. This phase requires the expression of a whole new set of genes. Maturation can also entail the expression of genes that create dormancy, a condition lost from most crop species.

The seed forms from the ovule after the egg cell within the ovule has been fertilized by a sperm cell from the pollen. As shown in detail for canola or oilseed rape, the resulting diploid zygote initially undergoes a singly mitotic cell division to produce two different cells: a small apical cell and a large basal cell. The smaller apical cell undergoes a series of cell divisions and differentiates into the embryo. The larger basal cell forms a short temporary structure, the suspensor, that serves as a conduit for nutrients into the growing embryo. In the dicot legumes and oilseeds such as canola, the two cotyledons are the prominent structures in the mature embryo, but in the monocot cereals the single cotyledon is reduced to a shield-shaped embryonic structure, the scutellum, which also stores oils.

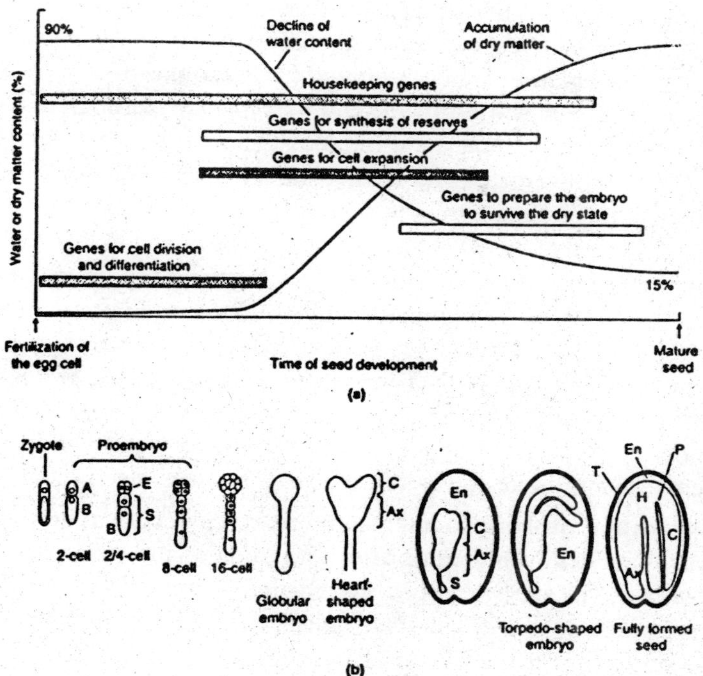

Figure 12.3: ***Events during the developmentof rape (canola) seed embryos and barley endosperm. (a) General scheme of the major events occurring during seed development. Dry weight increases as the storgae reserves (protein, starch, and oil) are laid down, displacing water in the storage tissues. A final decline inwater content to 12–15% occurs as the seed undergoes maturation drying. During development of the seed, different sets of genes come into play, as indicated by the horizontal bars. (b) Rape seed embryogenesis. After fertilization to form the zygote, mitosis occrs to produce an apical (A) and a basal (B) cell. the latter forms the suspensor (S), which facilitates development of the embryo (E) by being a conduit for mutrients but degenerates during the later stages of seed maturation. The apical cell undergoes extensive cell division and tissue formation until it forms a torpedo-shaped embryo embedded in a nutritive endosperm. Reserve deposition then occurs, with the endosperm and parent plant providing nutrients for the formation of oils and proteins inthe cotyledons. The supply of nutrients from the parent plant for reserve formation arrives via the phloem to the seed coat, from where the nutriens diffuse inward to the developing embryo. The endosperm is completely reabsorbed by the time the embryo hs completed reserve accumultion. A: apical Cell; Ax: axis; B: basal cell; C: cotyledons; E: embryo; En: endospenrm; H: hypocotyl; P: plumule (shoot apex); S: suspensor; T: testa (seed coat).***

The endosperm is the product of fusion between a second sperm cell from the pollen grain and a special ovule cell with two nuclei. This fusion product has a triploid (3*n*) chromosome complement. This unique cell forms the entire endosperm through many cell divisions. In legumes and oilseeds the endosperm develops but is often consumed again during the embryo's development. In cereals, the endosperm is the dominant feature of the mature seed. At the end of cereal seed development, most endosperm cells die and only a thin layer of starch-free cells around the periphery called the *aleurone layer* remains alive. After germination, these cells produce and secrete enzymes that digest the stored reserves to support seedling growth. The seed coat surrounds and protects the embryo (and endosperm, when present). In some species, including the cereals, the outermost covering layer is derived from maternal ovary tissue and is a fruit coat rather than a seed coat.

Deposition of the insoluble stored reserves occurs in storage parenchyma cells and begins after formation of the embryo and endosperm. The storage tissues expand greatly, and the dry weight of the seed increases as reserves replace water in the cells. In cereals, starch and protein in the endosperm account for about 75% and 12% of the mature dry weight of the seed, respectively, whereas the oil deposited in the scutellum is usually less than 5% of the dry weight. There are predominantly starch-storing legumes (dry beans, peas, chickpeas) and oil-storing (soybean, peanut, or groundnut) legumes; mature soybean seeds contain approximately equal amounts of starch and oil. Protein content of legume seeds is usually 25-40% of dry weight, with starch comprising about 40-55%, when present, or oil at 25% (soybean) to 50% (peanut). Sunflower and rape seeds contain 40-45% oil on a dry weight basis, and about 25% protein.

Starch is synthesized and stored as starch granules contained within amyloplasts, special starch-storing plastids. In cereals mutants of maize, there are defects in the biosynthetic pathway for starch, leading to the accumulation of sucrose and accounting for the sweetness of their seeds, commonly known as *sweet corn*. Protein is deposited within discrete protein bodies in the storage cells. In cereals, the proteins (prolamins) are nutritionally deficient in the essential amino acid lysine. Oats, in contrast, are rich in other proteins and are a better source of dietary protein than other cereals. Legume seeds contain proteins (globulins), which are deficient in the essential sulfur-containing amino acids, cysteine and methionine.

Seed oils, or triacylgylcerols, are deposited within small oil bodies. Species differ in the types of fatty acids (length, and degree of unsaturation) contained in their oils. For cooking and human health, the presence of unsaturated fatty acids (those containing one or more double bonds in the carbon chain, such as oleic and linoleic acids) is a desirable feature.

Seeds may also store antinutritional compounds. Phytic acid, an insoluble phosphate derivative, frequently is a minor component in seeds (1% dry weight or less), but it can bind essential dietary minerals in the intestines, making them unfavourable for absorption. Processing and preparation of many foods for human from seed may remove most of the phytic acid, but its presence in animal feed slows livestock growth. In addition, the phosphorus contained in phytic acid is not readily digested, so animal feeds are supplemented with phosphate, which can end up polluting waterways.

Some seeds, particularly those of legumes, may also contain enzymes inhibitors to protect them from predation by insects; these inhibitors can reduce digestibility of uncooked seeds. These and other proteins can evoke severe allergenic responses in susceptible people; for example, gliadins and glutens in wheat, globulins in legumes and an albumin storage protein in peanut.

Many variables affect seed set (the initial formation of the embryo and first growth of the seed), seed growth rate, and duration of storage reserve accumulation. Environmental stress at the time of pollination and fertilization can cause many seeds to abort, reducing the potential yield. Inherent genetic differences can regulate the number of cells in cotyledon or endosperm storage tissues, but the environment during seed development can modify the supply of assimilates available for storage product synthesis. Water deficit, temperature, light, plant competition, and nutrient availability are all variable affecting seed yield. High temperatures during seed filling can shorten the filling period, reducing seed size and yield. Water deficit can induce the same effects, and in some species it accelerates leaf senescence, prematurely terminating seed filling.

SEED MATURATION

The third phase of seed development is maturation and entry into quiescence. During the seed-filling phase, storage reserves replace water in the cell vacuoles, reducing seed water content to around 40-50%. During maturation, the seed synthesizes special sugars such as raffinose and a number of unusual proteins that bind water molecules. These may help protect cellular membranes and proteins and enable the cells to survive the final desiccation phase, when seed water content drops rapidly to less than 15%. At this low water content, biochemical processes essentially stop, allowing the seeds to persist for long periods in the dry state. Many maturation processes are controlled by the hormone abscisic acid (ABA), and mutant plants that do not make ABA or have a defect in gene regulation by ABA are viviparous: They germinate while still attached to the parent plant. After seeds are shed and fall to the ground, they imbibe water and, if dormant, they can survive in the soil for years.

Maturation and loss of water are essential in most crops, to let seeds be harvested and stored in the dry state. However, maintenance of seeds in the dry state does not deter insects from completing their life cycle in the stored seeds, a problem that is specially acute in developing countries, where people often seeds in granaries rather than in hermetically closed silos where they can be fumigated with insecticidal gases. The insects that damage stored seeds and their products (such as flour) are referred to as *stored product pests*. Prominent among these are the bruchids, beetles that lay eggs on pods in the field or on dry harvested seeds. The young larvae then burrow into the seed and continue to develop as they consume the seed's content. When the adults emerge after about a month, they leaves holes in the seeds. They then mate, and each female lays about 75 eggs on the seeds, thus starting a new life cycle. Low-resource farmers use various approaches to try to kill as many larvae as possible, such as spreading the seeds out in the hot sun.

Figure 12.4: ***Protecting seeds from damage caused by the larvae of seed weevils. When seeds are harvested they may contain larvae of seed weevils, insects that propagate when the seeds are in storage. Spreading the seeds out in the hot sun before storae in granaries can kill most larvae (See Figure 12.1).***

Seed drying also plays an important physiological role, because it signals to the seed that developmental processes are concluded. Thereafter, when the dry seed takes up water, its metabolism is now geared exclusively toward germination and seedling growth.

In many wild plant species, seed maturation is accompanied by the imposition of dormancy, a condition that also has a genetic basis. The term *dormancy* refers to the ability of seeds to delay germination until the environmental conditions are appropriate for survival of the seedlings. Dormancy is beneficial for wild species by timing seedling emergence to occur in the proper season, distributing seed germination over many years, and reducing competition among offspring. In crop species, seed dormancy is usually undesirable, because it leads to unpredictable germination, uneven crop growth, and variable yields. When early farmers began to harvest and store seeds for replanting, seeds that germinated readily after the rainy season and grew to maturity during the dry season were collected preferentially, resulting in a strong selection against dormancy. Thus, most domesticated crops show relatively little seed dormancy.

Nevertheless, a limited amount of dormancy is desirable in some crops. If rain and high humidity occur prior to harvest, cereal grains,

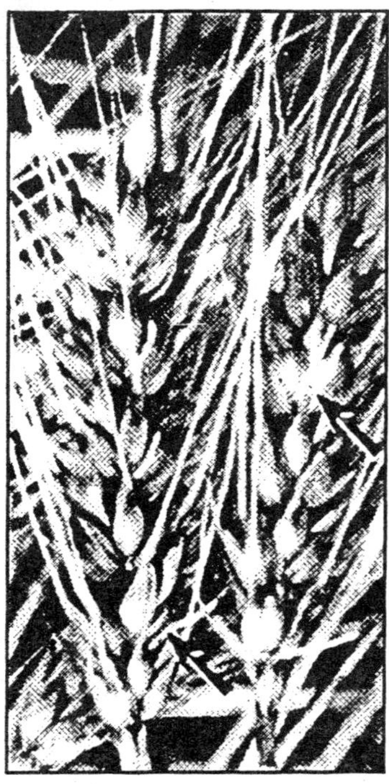

Figure 12.5: ***Precocious germination or pre-harvest sprouting of wheat. When wheat is left on the field too long in a wet summer, it may begin to sprout)arrowheads) before it is harvested. Sprouting occurs because domesticated wheat has lost its dormancy, still found in wild grasses. Flour made from such sprouted wheat produces poor quality bread and flour mills do not pay full price for such grain if they accept it at all.***

especially wheat and barley grown in temperate climate, are prone to germinate while still on the parent plant, a phenomenon known as *preharvest sprouting*. The result is poor-quality grain with unacceptable milling and malting properties and poor baking quality of the flour, causing losses to farmers running into the hundreds of millions of dollars. The retention of limited dormancy is therefore desirable in these cereals, preferable with dormancy being lost from mature dry grain while it is stored in preparation for planting the following spring. Loss of dormancy during dry storage (called *after-ripening*) is common for seeds of many species.

SEED GERMINATION

When farmers talk about seed germination, they mean the emergence of the seeing above the soil, but botanically speaking, germination starts when the dry seed begins to take up water and is completed when the embryo elongates and the embryonic root pushes its way through the seed or fruit coat. Imbibition, or the uptake of water, is a strictly physical phenomenon, much like water uptake by a sponge. Emergence of the seedlings, also called *stand establishment*, is caused by the growth of the embryo, supported by the reserves stored in the storage organs. After imbibition, metabolism resumes very quickly, initially using enzymes and reserves synthesized during development and conserved in the dry seed. Soon new genes are transcribed, and enzymes and new cellular components are made.

Respiration during germination uses stored sucrose and a series of modified sucrose compounds that contain galactose and are called the *raffinose series oligosaccharides* (*RSOs*). RSOs which occur at high levels in seeds of legumes, have attracted some interest in relation to human and monogastric farm animal nutrition because they escape digestion in the upper intestinal tract. Instead, they degrade and ferment in the colon, leading to the production of H_2 and CO_2, causing flatulence. The common bean is a notorious source of flatulence. Eliminating RSOs from soybeans improves efficiency of energy conversion when farmers use such soybean lines as feed for monogastric animals. Genetic engineers are now seeking ways to eliminate RSOs by disrupting their biosynthesis.

Postgermination events, such as the utilization of the stored polysaccharides, proteins, and oils and their conversion to small molecules such as sucrose and amino acids that can be transported, are quite well understood. In cereal grains, the living aleurone layer produces the digestive enzymes for starch in response to the hormone gibberellic acid released from the embryo. These enzymes are then secreted into the storage cells of the starchy endosperm, which the aleurone layer surrounds. Malting of barley for producing beer involves acceleration of this process so that the sugars released through the action of digestive enzymes are available for subsequent fermentation (alcohol production by yeast) in the brewing process. The success of malting requires maximizing endosperm reserves modification while minimizing seedling growth, a process that would consume the sugars.

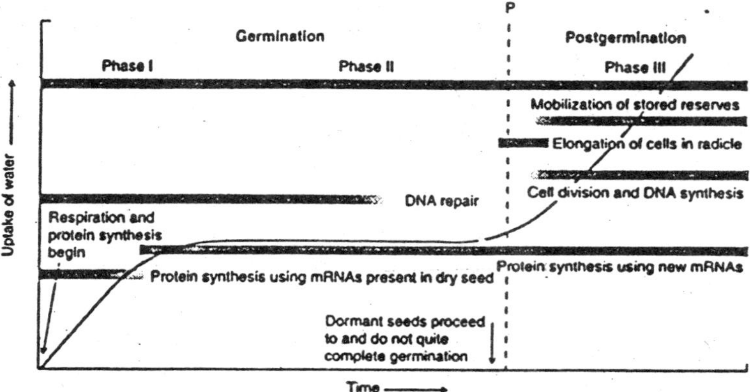

Figure 12.6: ***Some of the major events that occure during seed germination and subsequent postgerminative growth. Events are completed within hours or ater many weeks, depending upon the plant species and germination conditions. Dormant seeds complete all events up to the time of radicle emergence. The graphed line shows the increase infresh weight (water uptake). The extent to which germination proceeds during hydropriming, osmopriming, and matrix priming is marked with vertical line P.***

Mobilizing reserves in the cotyledons of legumes and oilseeds during seedling establishment involves enzymes produced within these tissues. As with the carbohydrates, the oils are converted to sucrose, which is transported to the growing seedling. The proteins are hydrolyzed into their 20 constituent amino acids, which are then converted to the amino acids asparagine and glutamine for transport.

Germination of (nondormant) crop seeds is affected mostly by soil temperature and moisture availability. Although many

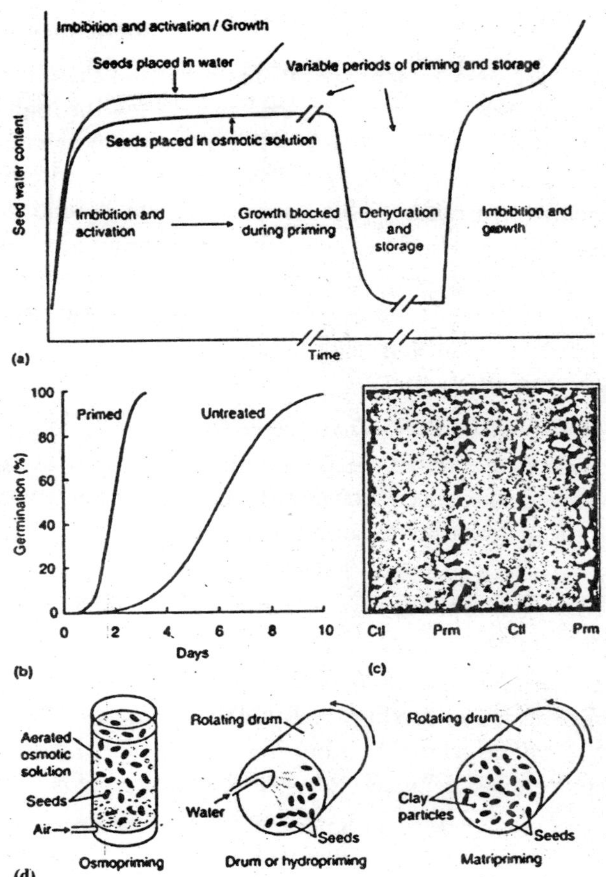

Figure 12.7: ***Seed priming can enhance seed germination. Seed priming involves allowing seeds to absorb enoughwater to initiate metablic processes, but insufficient water t complete germination (a). After the priming period, the seeds can be dried, stored, and transported to the former. When the seeds are imbibed again, the lag period before radicale emergence occurs is considerably reduced, improving the rate and uniformity of germination (b). for example, the emergence of muskmelon seedlings at a cool temperature (15°C) is improved by priming (c); Ctl = untreated control, Prm = primed). Priming can be achieved in several ways (d), including imbibition in an osmotic solution that controls water uptake by the seeds (osmopriming), slow addition of measured amounts of water to bring the seeds to a specific water content (drum or hydropriming) or combining seeds, clay particles, and water to allow the seeds to imbibe to a specific water content (matripriming).***

species and varieties can achieve satisfactory stand establishment in a wide range of environments, excessively low or high temperatures or water deficit can severely reduce the percentage of seeds that develop into plants. For some species seed priming can enhance seed performance. This technology, which controls water uptake by seeds, allows many early germination processes—but not radicle emergence—to be completed before sowing. Seed priming can thus lead to increased germination percentage and rate, germination under a broader range of environments, and improved seedling vigour and growth.

Several priming technology can be used:

- *Osmopriming* soaks seeds in aerated solutions of certain dissolved substances such as polyethylene glycol, potassium nitrate, or mannitol to control imbibition.
- *Hydropriming* or *drum priming* introduces liquid water to seeds in controlled, precise amounts to achieve a desired level of hydration.
- *Matripriming* mimics the natural process of imbibition from soil particles by mixing seeds with solid materials (such as diatomaceous earth, clay, or vermiculite products) containing water to control water uptake by seeds.

After priming, seeds retain tolerance to desiccation and can be dried for packaging, transportation, and marketing. However, storage life of primed seeds is generally reduced, so they must be handled with more care than untreated seeds. Seed priming has been applied most successfully to vegetable and ornamental species—such as lettuce, tomatoes, peppers, carrots, onions, pansies, and impatiens, as well as some turfgrass species—where seed quantities are relatively small and seed value is high. For field crops such as maize, wheat, cotton, or soybean, where seed quantities are high and seed costs are comparatively low, the expense of priming and the difficulties of processing huge quantities of seed has thus far outweighed its benefits; it is simply cheaper to sow more seed to assure satisfactory plant stands in the field.

Other technologies can also enhance seed performance and improve crop establishment. Inoculating legume seeds with

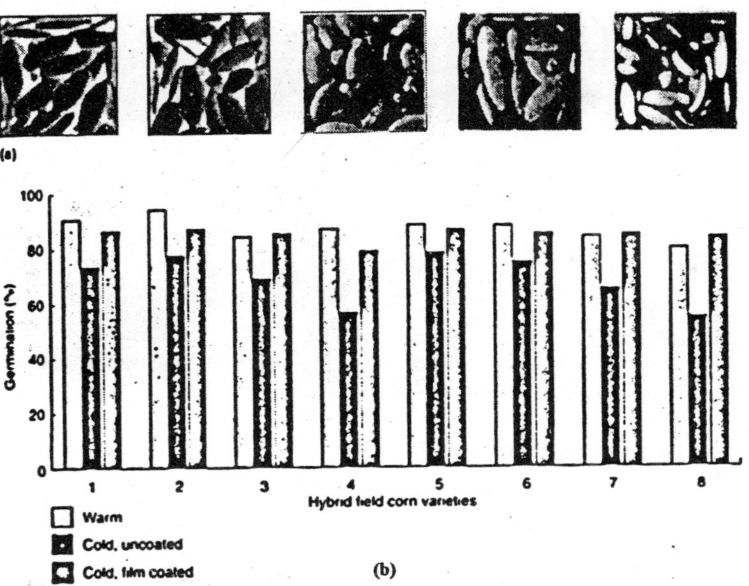

Figure 12.8: ***Film coating can be used to improve seed performance. Film coating applies a thin layer of a polymeric material to the outside of the seed (a). The polymer is applied as a mist as seeds fall though a special treatment chamber, ten is quickly dried. Protective chemicals such as fngicides included in the polymer are bound to the seed, allowing the use of less fungicide and reducing exposure of workers to the chemicals. Seeds treated with pesticides arerequired to becolored to prevent their accidental use as foods. Colors can also beused ot distinguis between varieties or treatments, as illustrated here with cucumber foods. Colors can also be used to distinguish between varietis or treatments, as illustrated here with cucmber seeds (a). Some coating materials, such as an organic starch-based polymer, also slow the rate of water uptake by seeds and prevent damage that can occur when dry seeds are planted in cool, wet soils (b).***

bacterial *Rhizobium* species has long been used to help establish a symbiotic nitrogen-fixing association with the growing plant. Seeds are often treated with fungicides or insecticides to protect seedling sagainst soil-borne diseases and insects. Recently, application of such chemicals to seeds is shifting from a dusting or slurry treatment to incorporating the chemicals into a polymeric film coating that adheres tightly to the seed. This prevents loss of the applied material from the coated seeds, reducing the amount

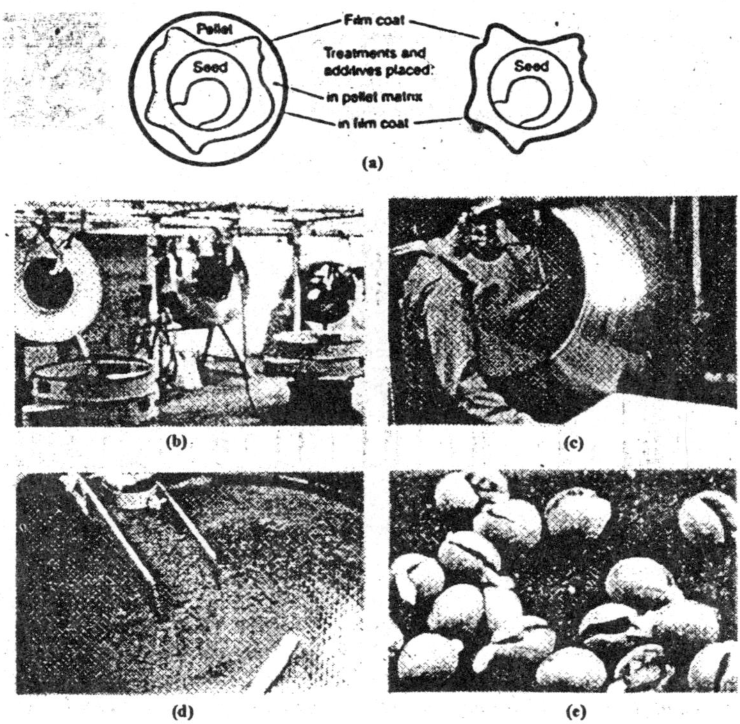

Figure 12.9: ***Seed pelleting allows irregularly shaped seeds tobeplanted by machines. Some seeds, as illustrated here by sugar beet (a), are irregularly shaped or are too small tobe easily planted with mechnical planters. In these cases the seeds can be made more uniforms in shape and increased in size by encasing them inside pellets of clay and othr materials. Pelleting is done in large rotating pans (b) in which the seeds are tumbling and being moistned with water or a binder. The peletin material is introduced a little at a time and gradually starts to form a ball around each seed (c). As the pellets near the desired size, they are screened and those that are too small are returned to the pan for additional coating (d). The finished pellets split open when the seeds imbibe, allowing the seedling to emerge, as illustrated here for lettuce seeds. (e). Film coats canbe applied directly to seeds or the pelleted seeds (see a).***

of chemical required and improving safety for people handling the seeds. Some coating materials, including a biodegrabale polymeric starch product, can slow water uptake by the seed and prevent the cellular damage that rapid imbibition can cause in cold soils. Other polymers have temperature-dependent water-permeability properties that prevent imbibition until a particular temperature is exceeded. This may allow coated seeds to be planted early in the fall or spring, because they undergo imbibition to start germination only when favourable temperature and moisture conditions have developed in the soil. Coating or pelleting are also use to make irregularly shaped seeds more uniform, making them more amenable for mechanical planting (for example, lettuce or sugar beet seeds).

SEED PRODUCTION IS OFTEN DISTINCT FROM CROP PRODUCTION

The production of seeds for planting can be essentially the same as producing the crop, or it can be a completely different commercial operation. For a number of the world's major crop, including wheat, rice, soybeans, and cotton, seeds are produced primarily using a self-pollinated or pure-line system. These crops do not outcross with other plants of the same species. Rather, each flower pollinates itself, resulting in seeds genetically identical to the parent plant. This has the advantage that once a superior variety has been developed, it can be propagated indefinitely by simply harvesting the seeds each year and replanting them.. Random mutations and a low incidence of outcrossing do occur, however, so vigilance is needed to maintain the integrity of the original variety.

For self-pollinated crops, quality assurance systems based on certification of planting stock pedigrees and field inspections for any off-type plants are efficient and effective. Because it is also possible for farmers to save their own seed of these types of crops for subsequent planting, people can widely distribute superior varieties quite easily. This system of developing superior varieties by breeding and then releasing the seeds to farmers through government-sponsored programs drove the Green Revolution advances in producing wheat and rice.

For some crops that normally outcross, or that can be readily manipulated to make crosses between parent lines, hybrid seeds are widely produced by seed companies and sold to farmers. To produce hybrid seeds, two inbred parent lines, each processing

(a)

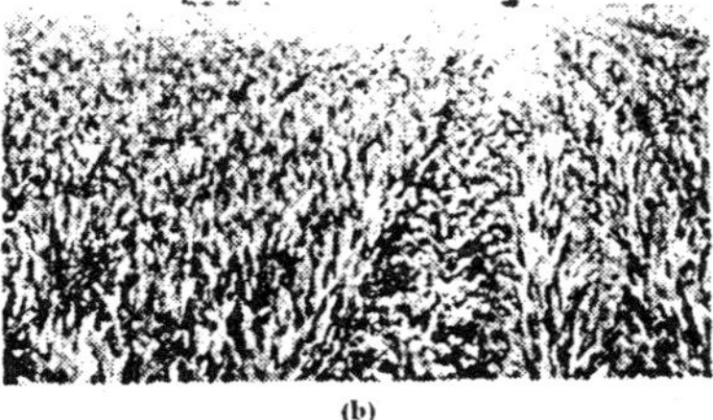

(b)

Figure 12.10: *Production fields for F_1 hybrid sunflower and maize seeds. (a) The two rows of multiflowered sunflower plants in the center are the male parents, and the single-flowered female parent plants are in the adjacent rows. The female flowers are male sterile (do not produce viable pollen) and are dependent on pollen carried by bees from the male plants for pollination and fertilization. After pollination, the male plants are removed, and only the female plants are harvested for the F_1 seeds. these will then be sold to farmers to produce the sunflower oilseed crop. In this case, the male parent also carries a gene that can overcome the sterility of the female parent, making the F_1 generation of plants fertile and allowing seed set. (b) F_1 hybrid seeds of maize are produced in a similar manner, except that the female rows are emasculated by removing the tassels (male flowers) by hand before they emerge. The single row in the center with the darker leaves is of the male parent, which soon produces tassels and sheds pollen to pollinate the ears (female flowers) onthe three adjacent female rows. In both sunflower and maize, note the distinctive plant types of the two parents.*

beneficial traits, are developed by repeated self-pollination. These parent lines are then crossed in a controlled way, with one line (male) pollinating the other (female) to produce F_1 hybrid seeds. Seeds produced on the females plants are then sold to farmers, who plant them to produce the commercial crop. The primary advantage of hybrid varieties is that when two distinct inbred (or genetically homozygous) parents are crossed, the offspring often grow much larger and more vigorously then either of the parent. This phenomenon, known as *hybrid vigor* or *heterosis*, has been exploited in many crops to improve yield and quality. Maize is the oldest example, but also sorghum, sunflowers, and in developed countries many vegetables, including tomatoes, peppers, carrots, cabbage, broccoli, melons, and onions, are grown almost exclusively from F_1 hybrid seeds.

Given the yield advantages of hybridization, efforts are under way toward developing hybrid in additional crops in which it is currently difficult to produce them. In many vegetables and flower species, pollination can be accomplished by hand, because the value of the resulting seeds justifies the additional expense. The labour-intensive production of hybrid seeds occurs primarily in China, Thailand, Chile, Central America, Kenya, and other countries where labour costs are comparatively low. In self-pollinated crops such as wheat, rice, cotton, or soybeans, hand pollination is not feasible and other methods must be employed. Methods to hybridize wheat or cotton by the use of chemicals to induce male sterility (failure of plants to produce viable pollen) have been developed but are not widely commercialized. However, much progress has been made in China with development of hybrid varieties of rice using a combination of genetic male sterility and labour-intensive methods. Hybrid varieties now represent 50% of the rice crop area in China and yield 15-20% more per hectare than conventional varieties. In addition, genetic engineers have developed new methods to control the fertility and pollination of crops to produce hybrid seeds, and are now introducing these into the market in crops such as the canola and maize. For many crops—including hybrids of all types, most vegetables, alfalfa, grasses. and other forages—seed production is a completely separate and specialized operation distinct from production of the marketed commodity. For example, field production of hybrid seeds requires interplanting two to eight rows of the female parent alternatively with one or two rows of the male parent. The female parent must be make sterile so that it can only be fertilized by pollen from the adjacent male rows. In maize, male sterility can easily be assured by removing the pollen-producing tassel (the male flower) at the top of the plant and leaving the ears (the female flowers) below.

Produing hybrid seeds is a challenge, because to succeed the producer must control many factors. For example, for pollen to be available when the female flowers are open and receptive, the two parent lines must flower at the same time. Management of pollination is also crucial for successful seed production. If the male flowers are not releasing pollen when the female flowers are receptive, then fertilization will not occur and no seeds will be produced. This difficulty is exacerbated by the fact that the best hybrids result when the parents are quite distinct and may have very different morphologies and growth rates.

In maize, wind carries the pollen to the adjacent female rows, so no special operations are required. In rice, even though genetically male-sterile female lines have been developed, the male flowers normally self-pollinate and do not readily release pollen to fertilize the female flowers. In China, the plant growth stimulator gibberellin is sprayed on the male plants to make them taller at the time pollen is shed, and workers drag ropes across the male rows to disperse the pollen so that the wind can carry it to the adjacent female rows. Many plants rely on insects to carry pollen among flowers. This is crucial in hybrid seed production. and large numbers of beehives are moved into seed fields during the flowering period to encourage pollination. However, bees forage widely (as much as 5 km) in search of nectar and pollen, and if crops of the same species are in the area, bees can carry pollen from other fields and contaminated the seed crop. Thus seed fields of insect-pollinated crops must be isolated by distances of 1-5 km from any crops of the same species to prevent genetic mixing by pollen transfer.

These example illustrate that seed production of modern agricultural varieties is a highly specialized operation conducted primarily by commercial breeding and seed production companies. In the developed countries, only 25-50% of a few crops (such as wheat and soybeans) are planted each year from farmer-saved seeds; farmers plant most crops from seeds they buy from seed companies each year. However, on a global basis, more than 83% of crops are planted from farmer-saved seeds. These seeds have the advantage that the plants are adapted to local conditions, but the disadvantage that they may not produce yields as high as those obtained with elite lines grown with the best management practices.

Expected yield increases are the primary reason why farmers pay companies for hybrid seeds. Farmers buy new seeds for every crop planting because self-harvested seeds will not "breed true". The plants from such seeds are not uniform in characteristics and yields. This follows from the laws of genetics and the independent assortment of chromosomes.

Hybridization is therefore a way to automatically protect the variety for the seed company, requiring the farmer to buy seeds from the supplier each year. Thus, in addition to yield advantages, the protection from unauthorized propagation that F_1 hybrid seeds offer has been a significant impetus to their commercial development.

New developments in biotechnology, however, may dramatically change this situation in the future. In some plants, seed development can occur without pollination and fertilization. In these plants, seeds naturally develop directly from cells of the ovary, so that the genetic composition of the seeds is identical to that of the parent. This phenomenon, known as *apomixis*, is essentially the same as when a potato is vegetatively propagated by a tuber or a fruit tree by grafting. If apomixis could be introduced in a controlled way into rice, for example, the F_1 hybrid would need to be created only once, and they that exact genetic combination could be transmitted to all subsequent generations.

SEED CERTIFICATION PROGRAMS

After the development of improved varieties by plant breeding, seeds must be maintained and propagated for distribution to farmers. If care is not exercised throughout the seed production process, genetic changes can occur over generations that reduce the quality of the variety. Random mutations, outcrossing with other varieties, or inadvertent mixtures with seeds of other varieties can result in contamination. This was often the fate of varieties released before the development of seed certification programs. These programs maintain a pedigree system that tracks each bag of seeds back to its parents over a limited number of generations. Breeder's seed is produced by the breeder and is the purest possible sample of seed representing the variety. This small amount of seed is then used to grow foundation seed, which in turn is used to produce registered seed, which is used to produce certified seed. The certified seed is sold to the farmers for producing a commercial crop. In this way, farmers are assumed that the seed they purchase is only a few generations away from the pure variety. The pedigree system allow the small amount of breeder's seed to be multiplied for commercial planting while maintaining the genetic purity of the variety.

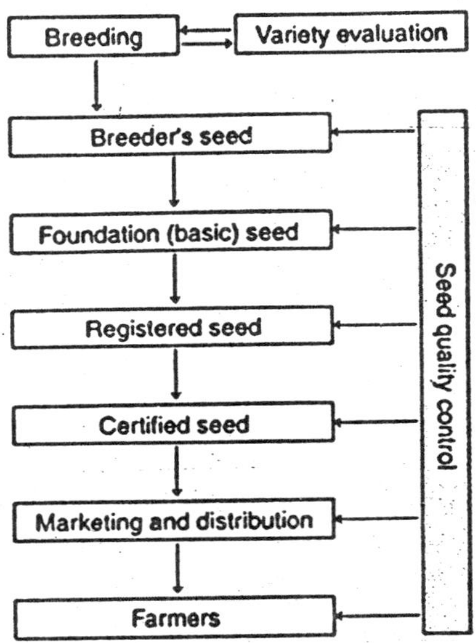

Figure 12.11: ***Flow chart for variety developmentand certified seed production. This diagram illustrates the essential features of a seed certification program from variety development through several multiplicatio steps before distribution to farmers. Quality standards (such as for isolattio distances, occurrence of off-types, and abosence of weeds) generations and become somewhat less stringent in later generations.***

In addition to maintaining the pedigree system, seed certification staff also inspect crops during growth for abnormal plants, diseases, weeds, and isolation (a minimum distance from plants of the same species that could outcross with plants in the seed field). Seeds from fields that do not meet the standards are rejected and cannot be sold as certified. Certified seeds must also meet standards of purity (freedom from weeds and foreign matter) and germination (viability). Together, these standards increase the benefits from improved varieties by maintaining genetic purity and providing high-quality seed for planting. Seed certification is most common in field crops such as cereals and legumes, although in many countries vegetable seed is also certified.

In developed and a number of developing countries, cultivar improvement, seed production, and seed certification programs are well established, and competition among commercial seed companies further ensures high standard for seed quality. However, many developing countries lack basic seed production, certification, and distribution systems, and most seeds are saved by farmers themselves or purchased locally. Without certifaction programs to maintain varietal purity, the advantages of miproved varieties introduced under these conditions can quickly disappear. Establishing seed production and certification programs is a high priority for international agricultural development, because most benefits of improved seed quality do not depend on farm size or level of technological sophistication.

Another important function of seed programs is to prevent the spread of diseases through contaminated seed. Some plant diseases are seed-borne, or capable of being carried on or in the seed. In many cases, even a very small percentage of contaminated seed is enough to cause a disease epidemic in the production field. For example, lettuce seed lots in California are rejected if only one seed in 30,000 carries lettuce mosaic virus. Because there is a global trade in agricultural seed, contaminated seed can introduce a new disease into a region where it has not existed before, creating destructive outbreaks and reducing crop yields. Many countries have phytosanitary regulations that strictly monitor the presence of diseases in seed production fields or on the seed, and obtaining a phytosanitary certificate is often required for the export of seed. The detection and monitoring of seed-borne disease has been greatly aided by the development of laboratory-based tests in which the pathogens are cultured from seed and identified or where they are detected by antibodies or DNA amplification techniques.

SEED BANKS

In addition to their roles in crop propagation and as foods, seeds are also a primary means of preserving crop germplasm. Most seeds can survive dehydration and remain viable for long periods if kept dry. The desiccation tolerance was an important evolutionary development for plants, allowing them to survive periods of adverse conditions such as dry seasons or freezing and to re-establish a plant population again when conditions were favourable. This cycle happens annually in many environments, or at much longer intervals in deserts, for example, where germination, plant growth, and seed dispersal may occur only once every few years or even decades in response to rare rainfall events. The ability to harvest, dry, and transport seeds that people could use to establish the next year's crop was fundamental to their adopting and expanding agriculture.

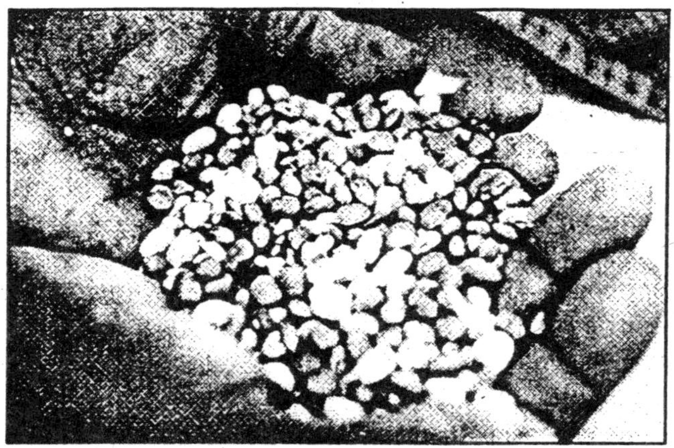

Figure 12.12: ***Landracs of cowpeas collected in a local market in Nigeria. The genetic diversity of the cowpeas grown by different farmers is reflected in the seed coat colors.***

To sustain the cropping cycle, seeds generally only need to be stored from one harvest to the next planting season—a period of nine months or less, in more cases. However, the natural ability of dry seeds to remain viable for much longer periods under proper storage conditions has allowed people to develop seed banks as a way to preserve the diversity of plant types important in agriculture. Early examples of seed banks were the All-Union Institute of Plant Industry in Russia, established by Nikolai Vavilov in the

1920s, and four regional plant introduction stations established in the United States in the 1940s. Currently, the U.S. National Plant Germplasm System consists of the National Center for Genetic Research Preservation in Fort Collins, Colorado, and 21 regional plant introduction stations where seeds and plant materials are preserved and propagated. There are now about 440,000 accessions, representing more than 38,000 species of plants, cataloged in the U.S. Germplasm Resources Information Network (GRIN), an electronic database of all genetic resources preserved by the system (www.ars-grin.gov/npgs/). Each year, the U.S. germplasm system alone distributes over 150,000 packets of seeds and other plant materials to more than 100 countries for use in crop improvement programs.

The introduction of high-yielding varieties of wheat and rice in many countries during the 1960s led to concern that traditional varieties, local landraces, and wild relatives of crop species would be lost if they were not collected and preserved. Landraces are not gentically uniform, a property reflected in the many colours of seeds found in local market of developing countries. These resources are crucial for maintaining the genetic diversity plant breeders require to continue to improve crop varieties. International initiatives to establish a global network of gene bands resulted in the establishment in 1974 of the International Board for Plant Genetic Resources (IBPGR, subsequently renamed IPGRI, International Plant Genetic Resources Institute) under the auspices of the Consultative Group on International Agricultural Research (CHIAR). This network has facilitated the collective and conservation of crop genetic resources and the establishment of germplasm repositories in developing countries. These efforts have focused primarily on crop plants, with the majority of accessions coming from the top 20 of the world's major crop species and their relatives. Because seeds are much easier and cheaper to store, of over 2.3 millions crop accessions held in bands around the world 97% are seeds and only 3% are plants or clonal materials.

More recently, attention has turned toward the remaining 99.99% or the world's plant species that are not crops and the growing concern about loss of global biodiversity. By the end of 2000 the Covention on Biological Diversity (CBD), developed following the Rio Earth Summit in 1992, had been ratified by 179 countries. The objectives of the CBD are the conservation of biological diversity, the sustainable use of biological resources, and the fair and equitable sharing of benefits arising from the use of genetic resources. The third objective recognizes that the vast majority of plant genetic resources originate in less developed countries, whereas these resources in crop improvement have been used primarily in the developed countries. Lobbying by the U.S. agricultural biotechnology industry because of concerns about limitations on its ability to continue to use these resources and retain ownership of varieties produced with those public resources has thus far prevented for United States government from ratifying the CBD. Other specific conservation initiatives include the Millennium Seed Bank Project of the Royal Botanic Gardens in Great Britain, which has the goal of collecting and preserving seeds of all of the United Kingdom's natvie seed-bearing plants and an additional 25,000 plant species from the tropical drylands by the year 2010.

Preserving seeds for the long term is not as simple as you might expect. The initial quality of the seeds before storage is important, and storage conditions are crucial for extending storage life. Most seed bank store-seeds in freezers at –20°C, but at the U.S. National Seed storage Laboratory, seeds are increasingly being stored in insulated tanks over liquid nitrogen (approximately –150°C). Depending on the species, seeds properly dried and stored at –20°C may retain viability for 20 to 50 years or more, whereas researchers expect those stored at –150°C to remain viable much longer. Researchers do not know how long, as experiments have been under way for only 30 years, and viability has changed little over this time. Experimental research on seed longevity in storage is a very slow process.

Collecting germplasm as seeds and preserving them in seed bands is *ex situ* conservation, in contrast to *in situ* conservation, which emphasizes preserving habitats and maintaining living populations of endangered species. Although *in situ* preservation might be preferable, because it maintains species or varieties as living plants that continue to adapt and create genetic diverstiy, in many cases it is not practical. Loss of habitat and changing crop production practices threaten many plants and crop varieties with extinction, and *ex situ* conservation may be the only way of preserving this germplasm. In addition, *ex situ* conservation complements *in situ* conservation by serving as an insurance policy against loss and by providing propagation material for *in situ* conservation programs.

Figure 12.13: ***Storage of seeds at low temperatures in seed banks.At th National Center for Genetic Resources Preservation in Fort Collins, Colorado, USA seeds are (a) dried and sealed in foil packets, then stord in large rooms held at –20°C, or (b) placed in plastic tubes and stored in insulated vessels over liquid nirogen at – 150°C. Seeds are expected t remain viable much longer at the lower temperatue, reducing the frequency at which they need to be grown into plants to produce new seeds.***

However, *ex situ* conservation is not simply putting seeds in storage. Because all seeds lose viability over time in storage, at some point the seeds must be planted to produce more seeds. Because the objective is to preserve as much genetic diverstiy as possible, regeneration is desirable when a high percentage of the seeds are still viable, rather than when only a few seeds are still able to germinate. Otherwise, the genetic contributions of all the dead seeds in the population will be lost. You can appreciate the magnitude of the challenge by considering that in the United States alone, over 400,000 seed accessions are in the collections. If the average lifetime in storage in 40 years (an optimistic assumption under standard conditions), then more than 10,000 accession must be grown each year just to replenish the existing stocks.

Thus far, we have considered only seeds that exhibit desiccation tolerance and can therefore be dried for storage. Such *orthodox* seeds represent about 90% of flowering plant species. In contrast, so-called *recalcitrant* seeds cannot tolerate dehydration and can generally by stored only for relatively short periods without either germinating or living viability. Many tropical plants produce recalcitrant seeds, as the temperature and moisture conditions are generally conducive to germination soon after the seeds are shed. However, many temperate trees and other plants also produce recalcitrant seeds. A number of important crops in tropical regions either produce recalcitrant seeds (rubber, oil palms, cocoa, mango, citrus, coconut, jackfruit, avocado) or do not produce seeds and are propagated vegetatively fiom cuttings or roots (bananas, cassava), making germplasm conservation much more labour intensive and costly.

SEEDS ARE THE MAJOR DELIVERY SYSTEM FOR PLANT BIOTECHNOLOGY.

Most crops are propagated by seeds, and seeds are therefore the primary delivery system by which breeders transfer the products of plant biotechnology to the farmer and then to the consumer. The unique traits adding value to a crop are sold to the farmer via the seeds, and the higher potential value is reflected in higher seed prices that enable the commercial developer to recoup the costs of creating the variety. Public institutions may not need to recoup their development costs when they release genetically engineered lines. The farmer expects a higher yield, a lower cost of production, or a higher price for the product because of the characteristics passed on through the genetically enhanced seeds. Thus genetic improvements increase seed value throughout the agricultural production and marketing chain.

"Value-added" traits developed through biotechnology (such as pest and disease resistance or improved nutritional quality) are most valuable if incorporated into elite varieties already highly adapted for yield and quality. The relatively limited number of sources for the best crop germplasm led biotechnology companies who developed the first genetically engineered seeds to acquire the leading seed companies, such as Du Pont's purchase of Pioneer Hi-Bred International (the world's largest seed company) and Monsanto's purchase of Holden Foundation Seeds and DeKalb Seeds. These and other companies in developed countries have also acquired seed companies in developing countries. Consolidation of seed companies by the corporations active in plant biotechnology research occured in the mid-1990s as the first products resulting from biotechnology began to reach the market. The high prices paid for these seed companies reflected the importance of introducing value-added traits into varieties that were already competitive in the marketplace.

Figure 12.14: ***Transgenic seeds ready for planting. GM seeds are more expensive than non-GM seeds because they contain value-added traits that save the farmer money. GM seeds can reduce the number of pesticid applications needed (for example, GM*** **Bt** ***cotton) or make weed control simpler and less expensive (for example, herbicide-resistant soybeans).***

In addition, as inputs such as protection from pests are increasingly delivered via improved seeds rather than as chemical pesticides, the value of seeds in the overall production system increases. A significant transfer of income from companies that make chemicals for crop protection (such as insecticides, fungicides, and so forth) to seed/biotechnology companies has already occured in the United States, Argentina, and other countries where crops engineered to resist pest have been introduced (although in many cases these represent separate division of the same company). For agronomic crops, where seeds have traditionally been a relatively low-cost component of the total production system, seed cost may increase considerably in the developed countries. As long as this is compensated by ever greater savings in other agronomic imputs, such as reduced insecticide or weed control costs, or by increases in market price of the product commodity, higher seed costs will be accepted. As has already occurred in high-value vegetable and ornamental crops, however, high seed costs lead to additional demands by the companies that creates them for intellectual property protection and by the farmers for enhanced seed performance in crop establishment.

Figure 12.15: ***Mickylee watermelon variety developed by the commerical vegetable breeding company Seminis Vegetable Seeds Inc. and protected by the Plant Variety Protection Act. This act procted plant varieties and the companies that created them before the more recent court rulings started protecting genetically improved plants using general utility patents.***

Legal mechanisms are in place protecting the intellectual property represented in genetically enhanced varieties. In the United States, the Plant Variety Protection Act (PVTA) provides a mechanism for registering seed-propagated varieties and restricting their sale. They can be sold only by the breeder or the company that developed them. Similar mechanisms are internationally recognized (for example, UPOV, the international Convention for the Protection of New Varieties of Plants). In addition, general utility patents can be awarded for novel plant varieties. These PVTA and patent rights give the developer of an improved crop variety the same type of protection that other inventors can receive for novel machines, pharmaceuticals, or processes. As discussed earlier, in self-pollinated crops it is difficult for a breeder to prevent unauthorized propagation and sale of a variety once it has been released. The situation is analogous to that in the publishing or music industries, where pirated copies of books or recording are often sold in places where copyright laws do not exist or are now enforced. Because farmers can but the improved variety once and then propagate their own seed thereafter, opportunities for economic return to seed companies are reduced and the companies have therefore been reluctant to invest heavily in breeding and genetic improvement of self-pollinated crops. In these crops, seeds have traditionally been a relatively low-cost input into the production system, with many varieties being developed by university and government-sponsored breeding programs and related to the agricultural industry with little or no royalty or patent protection. However, with the array of new traits that can be genetically incorporated and marketed through seeds, both public and private breeders developing those traits expect to receive a higher price for the seed and will enforce their intelluctual property rights. Thus, higher prices for seeds can be anticipated in these crops, and the issue of farmer-saved seed will need to be addressed.

This problem has already become an international issue, for example, with herbicide-tolerant soybeans. In the United States, farmers were required to sign "technology agreements" not to save or sell seed from crops planted with these varieties, and some compainies selling these seeds aggressively pursued violators of this agreement through lawsuits. At the same time, similar varieties were being sold in Argentina, where patent protection laws do not give the seed companies the ability to require the farmer to sign a restrictive technology agreement. Farmer-to-Farmer sales quickly distributed seeds of herbicide-tolerant (Roundup Ready®)

varieties, and within only a few years over 90% of the soybeans grown in Argentina carried this trait. As might be expected, U.S. farmers were angry that Argentine farmers were able to obtain the same varieties of soybeans with desired traits without paying the additional fees associated with the technology agreements, and the seed company's returns were reduced by the so-called brown-bag sales (off-market sales of seeds among farmers). With annual global seed exports estimated at US$3.6 billion and internal commercial seed markets totaling over six times this amount, considerable opportunity exists for profiting from avoiding laws protecting plant varieties. Inability to control propagation of self-pollinated crops deters seed companies from marketing improved varieties in countries where legal protection is not available or not enforced.

At the same time, some developments are so valuable to society that such constraints should not limit widespread distribution. On the one hand, for example, should rice having increased iron and β-carotene (*Golden* Rice™) be distributed only to those who can pay a licensing fee, when people who most need it probably cannot afford the fee? Fortunately, seeds of this nutritionally enhanced rice will be further developed and released through the International Rice Research Institute to make it freely available to people who need it. On the other hand, companies that have invested millions of dollars in developing improvements that benefit farmers or consumers of agricultural products also expect a reasonable return on investment. A two tiered system, where royalties are waived for countries when average farm incomes are below a given threshold, has been negotiated for *Golden* Rice ™, and has been proposed for general international adoption.

TERMINATOR GENE TECHNOLOGY

The conflict over control of seed distribution was sharply focused by the development and patenting of a method to genetically engineer seeds so that they would not be viable us propagules for producing a next generation. Developed by researchers at the U.S. Departments of Agriculture (USDA) in collaboration with the Delta & Pine Land Company, a major U.S. cotton seed company, this genetic "Technology Protection System" allows the distribution of seeds for planting that will germinate to establish the crop, but the seeds produced on those plants will not be viable for planting to produce a subsequent crop. This system was promptly termed the "terminator gene" by groups opposite to genetic engineering of crops, and a worldwide controversy erupted over it, even though it was and remain in the research stage and has not been transferred to commercial crops. Some countries prohibited its use, and various groups, including the Rockefeller Foundation, a major sponsor of the international agricultural research centers, criticized its intent. People expressed fear that poor farmers be forced to buy expensive seeds each year and that multinational companies would hold a monopoly on crop varieties. There was also concern about the potential spread of the nonviability trait to other nearby crops, making them useless as seed. The outcry was sufficient to cause Monsanto (which was in the process of acquiring Delta & Pine Land Company; that acquistion was not completed, however) to announce that it would not commercialize this technology. However, the USDA, Delta & Pine Land Company, and other agricultural biotechnology companies are continuing research on this and other methods to prevent unauthorized propagation of self-pollinated varieties.

The clear intent of this technology was to provide a mechanism for releasing self-pollinating varieties while affording the same protection from unauthorized propagation that is currently available only in hybrids. This would maintain profitable sales without competition from farmer-saved seed. However, this system would also prevent escape of genetically engineered varieties into the environment, because any seeds carrying the trait would not be viable. In addition, farmers in developed countries generally do not save seeds of most crops, either because they plant hybrids or because producing high-quality seed is difficult. In these cases, seed production has already been primarily in the hands of private companies for the past 50 years or more. Nonetheless, active competition still exists among seed companies selling those crops, including maize, sunflowers, and many vegetables, suggesting that limiting the ability of farmers to save their own seed does not automatically result in monopolistic seed companies. The reverse may be true: The opportunity for economic return increases investment and competition in crop breeding, as has been the case for hybrid varieties.

In countries where agricultural technology is less advanced, in contrast, the situation is quite different, because most farmers save their own seed or rely on government-sponsored seed distribution systems to provide seeds at low cost. The CGIAR, which sponsors a network of 16 international agricultural research centers, announced it would not release any germplasm incorporating technologies that prevent farmers from saving and replanting seeds. However, private companies developing genetically enhanced varieties may choose not to release them in countries lacking mechanisms (legal or biological) to prevent unauthorized propagation. The potential benefits of agricultural biotechnology, such as improving nutritional quality or reducing pesticide use, are too great to be withheld from most of the world's population because of intellectual property issues. As Norman Borlaug, the 1970 Nobel Peace Prize recipient noted, "This issue goes far beyond economics; it is also a matter for deep ethical consideration. Fundamentally, the issue is whether small-scale farmers of the developing world also have a right to share in the benefits of biotechnology".

SEED'S ABILITY TO FUNCTION AS A PROPAGULE

Low-resource farmers have traditionally selected their own seeds for planting the next season. This practice started disappearing in developed countries as public institutions and seed companies produced elite varieties that were either hybrids, as with maize, or were resistant to the most recently evolved pathogen strains, as with wheat. The situation of seed propagation by farmers becomes even more complex when a trait that is valuable in the crop interferes with the seed's ability to function as a propagule to establish that crop. This situation is arising more and more often as people change seeds by GM technology, but it is certainly not new. Consider seedless watermelons. Where do the seeds come from to plant a field of seedless watermelons? In this case researchers use a specific strategy of hybridization. The female parent is tetraploid, containing four copies of each chromosome (instead of the usual two copies), and the male parent is a normal diploid. When these plants make gametes, each egg cell has two copies of every chromosome and each pollen grain has one copy. The resulting zygote (fertilized egg cell) contains three copies of each chromosome. This does not stop the plant from growing, but problems arise when the flowers are forming and meiosis occurs. The odd number of chromosomes prevents pairing and proper distribution of the chromosomes to the daughter cells, resulting later in the fertilized egg cells in those plants failing to produce embryos and seeds. The fruits continue to develop, but the seeds abort at an early stage.

Figure 12.16: ***Collecting maize seeds for next season's planting in Colombia. Low-resource farmers collect their own seeds, but in both developed and developing contries commercial farmers obtain seeds from public institutions or seed companis for planting. The reason is that they want elite varieties with the best resistance genes orthe best hybrids. Plants grown from seeds produced by hybrids never produce the same high yield as the bybrids themselves.***

Some traits incorporated into varieties for their value in the end product can also impact how seeds are propagated and marketed. One example is the use of certain naturally occurring mutations that increase sugar contents in maize (sweet corn). Some of these mutations (such sa *shrunken*2 or *sh*2) limit the maize seed's ability to synthesize starch; instead, the seeds accumulate large amounts of sucrose. This is a valuable trait for sweet corn, because consumers prize the flavour, and the sweetness lasts longer after harvesting. However, because of the low starch content the seeds shrivel up when they dry out, hence the name "shrunken." The high sugar content makes the seeds attractive to insects and pathogens. These properties make it hard for the farmer to establish a crop in the field. Similarly, researchers have modified canola seeds to produce different types of oils, some of which reduce the ability of the seeds to germinate at lower temperatures. In soybean, eliminating raffinose to improve digestibility by animals may decrease seed longevity.

One application of the "terminator technology" discussed earlier is to make seeds non-viable and prevent unauthorized propagation. However, at a more basic level it represents the ability to control the seed generation in which a specific trait is expressed—or remains silent until a specific trigger activates it. Many potential applications of this technology to crops need not involve the protection of intellectual property. For example, in sweet corn, it would be possible to genetically engineer a normal *Sh*2 gene into *sh*2 mutant plants during the seed production generation. This would allow starch synthesis during seed development, resulting in higher-quality seed for the farmers. This normal *Sh*2 gene could then be silenced in the next generation, which produces the sweet corn for the consumer. Thus, although many condemned the terminator technology, people will find other useful applications. As is often the case in science, discoveries are neutral, but what people do with them is not.

Index

A

A. brevis 8
A. nuda 8
acephala 87
Actinidia deliciosa var. deliciosa 88
aerenchym 265
after-ripening 283
Agrobacterium 123, 231, 235
Agrobacterium tumefaciens 233
Agrostis stolonifera 88
aleurone layer 278
alleles 220
Allium cepa 87
americana 7
amylacea 7
Ananas comosus 87
apomixis 293
Arabidopsis 126, 161, 166, 237, 272
Arabidopsis thaliana 123, 252
Arabidposis betacea 88
Arachis hypogaea 87
ari-eGp 69
Asclepias curassavica 100
Asparagus officianalus 87
Astropa belladonna 87
AtHSF1 161
Avena fatua 8
Avena orientalis 8
Avena sativa 8, 87

B

B. oleracea 87
B.oleracea 87
Bacillus anthracis 204
Bacillus thuringiensis 36, 47, 98
barstar 270
Beta vulgaris 87, 88
Betula pendula 87
biennis 87
biotechnology 227
Brassica 121, 177
Brassica carinata 88
Brassica juncea 87
Brassica napus 87
Brassica nigra 87
Brassica oleracea 87
Brassica rapa 88

C

C1 126
calcium pumps 267
calciuın spike 267
Carica papaya 87
Castanea sativa 87
Castonea dentata 87
cell differentiation 254
cell elongation 254
chromoprotein 266
chromosomes 221
Chrysanthemum 87
Cichorium intybus 87
cis 121
Citrus sp. 87
cloning 258
comovirus 197
compactum 4
complementary strands 209
continuous cell division 254
Copia 66
Corynebacterium 118
Cotesia plutellae 100
Creptis acetylenics. 182
crown galls 233
CrtO 125
Cucumis melo 87
Cucurbita pepo 88
Cucurbita texana 88
Cyphomandra thaliana 88

D

Danaus plexippus 99
dapA 118
Daucus carotta 87
de facto 146
de novo 174
dehiscence 263
Dianthus carophyllatus 87
Dioscorephyllum cumminsii 127
distichon, hexastichon, agriocrithon, deficiens 8
dormancy 281
drum priming 286
duplicated 223
durum 54

E

E. coli 131, 173
E.coli 188, 203, 209
Echinochloa crus-galli 9
Eleusine coracana 9
Embryogenesis 252
Erwinia uredovora 123
Escherichia coli 118, 228
et al. 70, 78, 100
71, 72, 73, 75, 76, 77, 78, 79, 80, 81, 84
Eucalyptus camaldulensis 88
Eulophus pennicornis 99
Eustoma grandiflorum 88
everta 7
ex situ 298, 299
exons 224
expression profiling 237

F

Flowers and Fertilization 249
fruit 263
fruit ripening 263
fruit set 263
fruits 263

G

GE) 232
gene banks 275
gene cloning 228
gene of interest 232
genes 206, 219
genetic code 213
genetically enhanced 232
genetically modified 232
Gladiolus sp. 88
Golden 304
Gossypium hirsutum 87
Gramineae 1
gusA 161

H

Haematococcus pluvialis. 126
Haplopappus 220
Helianthus annuum 88
Helianthus tuberosus 127
heterosis 290
heterozygous 220
high dose 38
homozygous 220
Hordeum laevigatum 63, 68
Hordeum spontaneum 8, 79
Hordeum spontaneum, 79
Hordeum vulgare 7
Horedeum vulgare 87
HVAI; 21
hybrid vigor 290
Hydropriming 286

I

in planta 190
in situ 298, 299
in vitro 142
in vivo 100, 129
in-vitro 103
in-vivo 103
indica 7
intercalary meristems 244
internodes 240
introns 224
Ipomoea batatas 88

J

japonica 7
Juglans sp. 88

K

kirilowii 200
Klebsiella pneumoniae 172

L

Lacanobiua oleracea 99
Lactua sativa 88
Liguidambar. sp. 88

line up 223
Linum usitatissium 88
liquid endosperm 252
Lolium perenne 93
Lupinus angustifolius 88
Lycopersicon esculentum 88
lysCM4 118

M

Malus domestica 87
Matripriming 286
Medicago sativa 87, 126
meiosis 221
meristem culture 259
meristems 242
Mla 71
MlLa 68
morifolium 87
Mortierella alpina 182
multigenic 220
Mushi-manju 56
mycorrhizae 255
Mysus persicae 99

N

NGAAN 161
Nicotiana benthamiana 88, 200
Nicotiana tabacum 88
Nicotianum tobacum 98
nodes 256
nuclear genome 211
nucleases 265

O

orthodox 300
Oryza glaberrima 7
Oryza rufipogon 7
Oryza sativa 7, 87
Osmopriming 286

P

Panicum miliaceum 9
Pelargonium sp. 87
Pennisetum glaucum 9
peptide bonds 211
pericycle 244
photoperiodism 248
phytochromes 266
Picea abies 88
Picea sp. 88
Pinus sp. 87
Pisum sativum 87
Plutella xylostella 100
Poa patensis 88
polypeptide 211
Populus 93
Populus sp. 87
praecox 7
preharvest sprouting 282
primordia 244
programmed cell death 264
protein-coding 224
Puccinia striiformis 81

Q

quantitative traits 220

R

R 126
raffinose series oligosaccharides 283
recalcitrant 300
refuge 38
regulatory regions 226
restiction enzymes 228
Rhizobium 287
Rosa hybrida 87
RSOs 283
Rubus idaeus 88
rym4 67, 81
rym5 67, 81

S

S.mutans 197
saccharata 7
Saccharum officinarum 88
Saintpaulia ionantha 87
Sake-manju 56
sdw 1 75
sdwl 69
Secale cereale 9
Secale montanum 9
seeds 263
selectable marker 232
semiconservative replication 210
separation 223
Setaria italica 9
setting 263
shrunken 307
Sinapsis alba 88
situ 70
Solanum melanogea 88
Solanum tuberosum 87
Sorghum bicolor 88
Sorghum vulgare 9
spelta 4
sphaero-coccum 4
spike 251
spikelets 259
stand establishment 283
stored product pests 280
Streptococcus mutans 194
sweet corn. 278

T

T-DNA 233
T. polonicum 4
T. turgidum 4
T.carthlicum 4
T.dicoccoides 4
T.monococcum 4
T.turanicum 4
Tagetes sp. 87
Ti plasmid 233
totipotency 230
trans 121
transcription factors 226
Trichosanthes 200
Triticeae 64, 82
Triticum aestivum 4, 5, 88
Triticum boeoticum 4

Triticum dicoccum 4
Triticum durum 4
Triticum monococcum 4
Triticum timopheevi 4

V

Vaccinium oxycoccus 88
vector 229
vegetables 263
vegetative propagation 258
vernalization 248
Vigna unguiculata 197
Vitis vinifera 88

W

Waugh et al. 65

Z

Zea mays 6, 87
Zea mexicana 7